DO ROBOTS MAKE LOVE?

DO ROBOTS MAKE LOVE?

FROM AI TO IMMORTALITY – UNDERSTANDING TRANSHUMANISM IN 12 QUESTIONS

Laurent Alexandre & Jean-Michel Besnier

With the collaboration of Nicolas Chevassus-au-Louis

An Hachette UK Company
www.hachette.co.uk

First published in Great Britain in 2018 by Cassell, an imprint of Octopus Publishing Group Ltd, Carmelite House, 50 Victoria Embankment, London EC4Y 0DZ
www.octopusbooks.co.uk
www.octopusbooksusa.com

Originally published in France as: *Les robots font-ils l'amour? Le transhumanisme en 12 questions* by Laurent ALEXANDRE & Jean-Michel BESNIER
© Dunod, Malakoff, 2016

Distributed in the US by Hachette Book Group, 1290 Avenue of the Americas, 4th and 5th Floors, New York, NY 10104

Distributed in Canada by Canadian Manda Group, 664 Annette St., Toronto, Ontario, Canada M6S 2C8

ISBN 978-1-78840-070-1

A CIP catalogue record for this book is available from the British Library.

Printed and bound by CPI Group (UK) Ltd, Croydon, CR0 4YY

10 9 8 7 6 5 4 3 2 1

Publishing Director: Trevor Davies
Senior Editor: Leanne Bryan
Junior Designer: Jack Storey
Translator: JMS Books LLP
Typesetter: Jeremy Tilston
Cover Illustration: Quibe
Production Manager: Peter Hunt

TABLE OF CONTENTS

PROLOGUE

The augmented human, synthetic biology, bionic prostheses, artificial intelligence: technological advances are mounting up at stupefying speed, and ideas that only a decade ago would have been the preserve of science fiction are now being actively researched in laboratories. Machines based on artificial intelligence are revealing the first glimpses of their extraordinary power; after the chess defeats inflicted upon Gary Kasparov by Deep Blue (designed by IBM in 1997), and Lee Sedol's loss at the game of Go to AlphaGo (invented by Google in 2016), the fields in which human intelligence still outstrips that of machines are shrinking.

The economic upheaval to be anticipated from all this is considerable, and it is impossible to list all the professions that will be turned upside down by the new wave of automation. Unlike the steam engines that became ubiquitous in industry during the 19th century and the robots that did the same in the second half of the 20th, these new machines are not replacing human strength. Instead, they are taking over part of what was once thought to be unique to humankind: knowledge, judgment, analysis and even rational thought.

This prodigious acceleration of technology has been expedited by the convergence of four disciplines that had hitherto evolved separately: nanotechnology, which manipulates matter at the atomic level; biotechnology, which models life itself; information technology (in its most fundamental aspects in particular); and lastly, cognitive science, which is based on the functioning of the human brain. It is this explosion of NBIC (nanotechnology, biotechnology, information technology and cognitive science) that has made it possible to imagine the unparalleled, unprecedented, Promethean project examined in this book: modifying human beings – improving them, augmenting them and surpassing them. For the highly influential transhumanists in Silicon Valley who are at the heart of the NBIC convergence revolution, this enhancement of the human species through technology represents the only chance for *Homo sapiens* to avoid being overtaken by the very machines they themselves invented. In fact, such man/machine hybrids have already made an appearance – we need think only of the artificial heart developed by Carmat that has been implanted in several patients suffering from cardiac failure. However, this is a mere precursor to what will be possible in a few decades: editing human DNA to eliminate the sequences responsible for genetic illnesses, using 3D printers to manufacture organs, transcranial magnetic stimulation, linking brain function to devices with artificial intelligence, amplifying perceptive faculties like physical strength. And even, for some, the prospect of extending life expectancy indefinitely, to a point when the demise of death itself can be imagined.

While such expectations might have transhumanists afire with enthusiasm, they have become a source of worry for other schools of thought. What will be left of the free will of humans who are inextricably interlinked with their machines? Is living to the age of a thousand really desirable? How will augmented humans coexist with others? Should we not fear a kind of bio-totalitarianism, like the one described in Aldous Huxley's *Brave New World*? At that time (1932), it was pure science fiction, but today's unease is based on a realistic anticipation of our possible futures.

Having had the opportunity to debate these matters on many occasions, crossing intellectual swords and exchanging arguments, we cannot agree on answers to these questions.

There's no getting around it – our disagreements will remain deep-seated. However, during such debates, we have managed to establish that our positions converge on two points that may be even more fundamental: the importance of discussion that is reasoned, well argued and respectful of the other point of view, and the conviction that technology is neither good nor bad per se and that everything will depend on the use humanity chooses to make of it.

It was this discovery that prompted us to write this book as a dialogue, so the reader should not expect to find some ultimate reconciliation or a sudden ecumenical consensus. This book is a conflict, a robust debate, an adversarial dispute, much like those practised by the Greeks for the greater good of their democracy. Our wider hope is that our exchanges of opinions will similarly invigorate popular discussion of the enormous challenges that NBIC presents to our very humanity.

Laurent Alexandre and Jean-Michel Besnier

1
SHOULD THE HUMAN
RACE BE IMPROVED?

Man is becoming the overseer of Creation, an inventor of phenomena; and in this respect, we shall never be able to circumscribe the power that he may acquire over Nature through future progress in the experimental sciences.

Claude Bernard, 1865

A technological revolution is under way as nanotechnology, biotechnology and artificial intelligence converge, making improvements to the performance of the human body and brain a real possibility. Technology can already create an augmented human and will be able to achieve more and more in the future. But should it take such a step?

Laurent Alexandre: The role of technology is to ensure good standards of living and to improve the conditions of human life. No one is against progress in medicine, which has led to ever-greater improvements in life expectancy, and such advances will continue. There are plenty of reasons for accepting the correction of biological weaknesses when technology permits. Take disorders affecting the retina, for example: the most common cause of sight loss in the developed world is age-related macular degeneration (AMD), which results in blindness as the centre of the retina gradually degrades. The number of people suffering from AMD in the entire world is projected to be 196 million by 2020 and 288 million by 2040 as the population ages. The condition already affects more than 600,000 people in the UK, over 1 million people in France and about 11 million Americans, and these figures are set to more than double by 2050. In addition to AMD there are several types of retinal pathology that also inexorably result in blindness for which there is no effective treatment. Progress in electronics (such as in biotechnology) will help us deal with AMD and other serious conditions more and more effectively. Why deprive ourselves of such techniques?

Technological promise in combatting blindness

There are two branches of technology that hold some promise of treatments for age-related macular degeneration (AMD). The first of these involves placing electronic implants in the retina or directly in the cerebral cortex with a connection to a micro-camera; it is the logical next step after treatment for deafness with cochlear implants. This bionic eye currently provides the patient with only poor sight, but the constant progress being made in microprocessors and electronic image capture provides hope that implants with multiples of tens of thousands of pixels, bringing real comfort of vision, might be achieved before 2025. The second branch involves biological technologies such as stem cells and gene therapy. In April 2011, a Japanese team announced in the journal *Nature* that they had manufactured mouse retinas from embryonic stem cells grown in a test tube; the use of stem cells for disorders of the human retina should begin by around 2025. Gene therapy similarly offers hope for young patients affected by hereditary retinitis. The first gene therapy used on retinitis pigmentosa in dogs has achieved a normalization of retinal function beyond all expectations, and the transfer of such results to humans is already under way. Experimental gene therapy published at the start of 2012 enabled the partial restoration of vision in three patients affected by a form of Leber congenital amaurosis. This rare condition involves an incurable degeneration of the receptors in the retina, resulting in total blindness before the age of 30. LA

Jean-Michel Besnier: In fact, it is not a question of depriving ourselves of these techniques. Indeed, must we accept everything of which we are capable? The physicist Dennis Gabor, the inventor of holography (which won him the Nobel Prize for Physics in 1971), said: "Everything that is technically practicable deserves to be realised, whatever the ethical cost." It's easy to get agitated about the cynicism implicit in this axiom, but it has the power of a natural law among the cheerleaders for the omnipotent market, who are convinced that developments in technologies will obey the same mechanisms as the natural selection of species. We try to do what is ethically right these days, of course, with committees examining the acceptability of technological breakthroughs, but it is very challenging, as encouraging innovation at any price has become a real article of faith among political and industrial decision-makers.

Laurent: I don't agree with you at all on this point, as I support a culture that promotes innovation. We go further because we can. In future, a distinction will no longer be drawn between a human who has been repaired and one who has been augmented. If blind people in 2080 wish to receive an implant of an artificial retina that gives them better-than-normal vision, are we going to throw them in prison? The answer has to be no! Over the next few decades, we shall be moving on from healthcare that repairs to healthcare that improves. Let's not forget that a vaccinated human is already an augmented human!

Jean-Michel: Let me pick up on this notion of "healthcare that improves". The only definition of a human being that is to me beyond dispute is the one offered by Jean-Jacques Rousseau: Man is a perfectible being ("*l'homme est un être perfectible*"). In other words, he is destined to improve himself endlessly, having been born unfinished. His first years impose upon him the need to tear himself away from the inertia displayed by animals, which, at birth, are fully and immediately what they will be at the moment of death.

Laurent: We have gone beyond Rousseau's definition: animals can be augmented as well, thanks to the progress made in biotechnology. Recent studies are taking us closer to Pierre Boulle's novel *Planet of the Apes* (1963). Three experiments, the findings from the most recent of which were published in the journal *Current Biology* on 19 February 2015, have enabled us to enhance the mental capacity of mice by injecting them with human brain cells, thereby modifying their DNA with segments of human chromosomes. The potential consequences are mind-boggling; how are you going to stop dog lovers from ordering a pet that is more intelligent, more empathetic, more "human"? There will always be people sympathetic to demands for improving animal cognition. Society will be presented with a *fait accompli*, as it is nowadays with children born to homosexual couples through the intervention of third-party surrogate mothers. In the name of what morality can we forbid chimpanzees from being more intelligent in the future? With the dignity of animals and respect for other species gaining greater and greater acceptance as concepts, how should we treat animals when they attain IQs approaching those of humans?

Jean-Michel: The line of argument you are developing there clearly shows how "improving the species" – every species – has become an obsession, although it was disdained for a long time per se by premodern societies that preferred to respect traditions or higher powers. However, I reject such a position as archaism or conservatism. We are modern people and we are modern because we continue to think that tomorrow could and should be better than yesterday. In this sense, technophilia is part and parcel of our nature, and I defy the Luddites and dwindling numbers of others suspicious of technology to suggest otherwise. There is also the consideration that the concept of improving the species is known by another name, which has been compromised by history: eugenics.

Laurent: We have jumped on the eugenics bandwagon without realizing it. Down's syndrome is disappearing before our eyes; in the UK about 90 percent of pregnancies with a prenatal diagnosis of Down's syndrome are terminated, in France it is about 97 percent, and in the US the figure is estimated to be 67–85 percent. There is strong social pressure to "eradicate" the disorder from society, and few people receiving a Down's syndrome diagnosis for their unborn child are able to resist this pressure; certainly I would not want to belittle these decisions. However, up until now, genetic techniques have been able to detect only a handful of pathologies, whereas in the future, sequencing a baby's DNA – the three billion chemical messengers that make up its genetic identity – will be a game-changer. It is already possible to make a complete genomic diagnosis of an embryo from a simple blood sample taken from the pregnant mother; there's no longer any need to take amniotic fluid via amniocentesis. Thus, one of the last obstacles to routine prenatal diagnosis – fear of miscarriage, which would occur in 0.5–1 percent of cases after amniocentesis – has been eliminated. Furthermore, powerful algorithms allow us to differentiate the DNA sequences of the unborn baby and the mother. As the cost of DNA sequencing plummets, this technique will become routine before 2025. It will be possible to systematically identify during pregnancy thousands of potential illnesses without endangering either the unborn baby or the mother. We have more or less eradicated Down's syndrome in 30 years, even though people with this disorder are amiable, have a normal life expectancy and do not suffer in any way. Why would we act differently with other conditions in the future? From a political perspective, how can parents be prevented from preferring to have beautiful, gifted children when abortion on demand is freely available, whatever the condition of the embryo, and when abortion for intellectual disability (principally Down's syndrome) is legal, socially acceptable and encouraged by the powers that be? Parents will soon be offered the dream of a child designed à la carte. If prenatal diagnosis allows us to "prevent the worst" by eliminating foetuses that reveal deformities, then pre-implantation diagnosis amounts to "selecting the best" – embryos obtained via in vitro fertilization are already sorted. There will be a strong uptake among parents once the final secondary effects of in vitro fertilization are brought under control, and it will be less distressing from a moral point of view to eliminate a foetus in a test tube than in the womb. The return of eugenics is a political bombshell that has gone unnoticed, and I am extremely sorry about that.

Jean-Michel: And it's exactly this that we can discuss as the "modern people" we both are, more or less on board with the cause of technology. Should we accept the "improvement" obsession to the point of agreeing with the notion of "enhancement" as advocated by the transhumanists? What is the difference between the French biologist Jean Rostand expressing his enthusiasm for the view that biology is making the dream of alchemists and visionaries come true – the dream of transforming Man – and the American transhumanist Ray Kurzweil's euphoria in announcing the impending advent of a posthuman who will free us from ourselves and our as yet too limited perfectibility? What might Kurzweil, the transhumanist, read in the biologist Rostand's book *Aux frontières du surhumain* (which translates as "At the Frontiers of the Superhuman")? And what would the biologist have welcomed in the prophecies formulated by the transhumanist in his book *The Singularity is Near: When Humans Transcend Biology*?

Rostand and Kurzweil

In *Aux frontières du surhumain* (1962), Jean Rostand explains how "the most audacious gamble in human history, transforming Man himself" is "a metamorphosis that may be closer than one might think". In *The Singularity is Near: When Humans Transcend Biology* (2005), Ray Kurzweil enthusiastically describes what he calls "the technological singularity", born of the marriage of biotechnology, robotics and artificial intelligence. The two books demonstrate the same confidence in the power of technology but are very different at heart. In Rostand, for example, there is a constant preoccupation with sharing the benefits of scientific research, while Kurzweil posits a fairly aggressive individualism. Essentially, neither is talking about the same thing – the "eugenics" of the former cannot exist without a concern for truth and consistency, and the latter insists only on efficiency and disruption. JMB

Laurent: *The Singularity is Near* has been called "the Singularitarian bible". The emergence of new biological creatures or intelligent electronics will have religious consequences: some theologians, such as Christopher J Benek, a Florida pastor who advocates Christian transhumanism, has expressed the desire that machines gifted with intelligence should receive baptism if they so wish. The confluence of nanotechnology, biotechnology, information technology and cognitive science (NBIC) poses unprecedented questions for the future of humanity. The 21st century will be anything but a long, quiet river!

Jean-Michel: Definitely. In order to prepare for it, I can only restate my support for the "humanogenetic" role that has fallen to technology, but I also demand that the symbolic dimension proper to the human species be preserved. Technology (tools) and language (words) have been rightly identified by paleoanthropologists and philosophers as the indices by which our species, and our species alone, can claim to have a history. And I am indeed saying technology *and* language. The Classical Greek philosopher Plato emphasizes this in the dialogue *Protagoras*, when he concludes his retelling of the myth of Prometheus: if Man had not been given fire and a knowledge of art and technology on his journey through life, he would not have survived. Society would have been wilful and chaotic, undermined by chance and selfishness, and would have eventually proved impossible to bear. Having foreseen this, Zeus charged Hermes with conveying the art of politics to humanity – using language to construct arguments, to consider and to decide the directions to be taken as a result of the opportunities offered by technology, with a view to creating a harmonious society where good living would be guaranteed. The position that I assume, which does not consign me to the ranks of the technophobes, is untimely in an era of adoration for numbers and calculations. The challenge offered by digitalization to considered argument is emblematic of the evolution of technological sciences, which are wiping out languages and culture and at best retaining only the signs and codes that they require. The position I propose should at least make it possible to escape the abstraction of a face-off between bio-conservatism and technological progressivism.

Laurent: Such a face-off already exists, however! The French are ultra bioconservative: only 13 percent are in favour of increasing a child's IQ by treating the foetus, compared with 38 percent of Indians and 39 percent of Chinese. Among young Chinese, the percentage increases to 50 percent. The Chinese are, in fact, the most permissive society as far as technology is concerned and would have few misgivings about increasing their children's IQs using biotechnological methods. The first genetic manipulation of 86 human embryos has already been carried out by Chinese scientists, in April 2016 – they published their findings immediately after an international petition opposed to such experiments hit the headlines! When the technology is ready, a country where a consensus on the mental enhancement of children holds sway could obtain a considerable geopolitical advantage in a knowledge society. The Swedish philosopher Nick Bostrom, of Oxford University, has estimated that selecting embryos after sequencing could enhance the IQ of a country's population by 60 points within several decades. Inclusion of genetic manipulation of the embryos could achieve an even more spectacular improvement. The eugenist countries would rapidly become the masters!

Jean-Michel: This fetishization of IQ, which I feel comes from another era, is completely alien to me. The understanding of intelligence that it implies is so narrow that it cannot be defended without recourse to the past. In my opinion, a desirable improvement of the human race is not the enhancement of the physical prowess and mental faculties of individuals that you describe – unless you want to reduce a human being to an animal or a machine by imposing measurements upon him, or reducing him to the algorithms and metabolisms so beloved of digital culture and its "GAFA" henchmen: Google, Apple, Facebook and Amazon.

Laurent: But whether you like it or not, GAFA will define the shapes and contours of the humanity of tomorrow, which may even include transforming the idea of what it is to be human.

Jean-Michel: It's no secret that I disagree – but discussing the reasons for this is exactly the point of this book.

2
WILL THE HUMAN RACE HAVE TO CHANGE THE WAY IT REPRODUCES?

The Bible condemns women to endure the pain of childbirth ("in sorrow thou shalt bring forth children") and they have submitted to this suffering for millennia. But is it really unavoidable? Technological progress means we can look forward to the reality of artificial wombs, making it possible to incubate future humans in vitro.

Laurent Alexandre: Let's start by acknowledging one thing. We humans have already changed the way we reproduce. Family planning has revolutionized the position of women and the way the family is organized, and we are only going to accelerate this process because most parents harbour a desire for a perfect child and because society encourages a minimization of risk during pregnancy. Techno-maternity will assert itself; childbirth at home, with no medical back-up and no epidural, which seems insane nowadays, was the norm in the 1930s. What's more, screening embryos and eliminating anomalous foetuses will become routine stages in any sensible pregnancy.

Jean-Michel Besnier: The logical conclusion of all this is essentially ectogenesis, i.e. incubation of the foetus outside the mother's body. The idea seems to smack rather of the dystopia of Aldous Huxley's *Brave New World*. But ectogenesis has already been employed with ewes and will be practicable for humans within a few decades. The biologist and philosopher Henri Atlan has posited 2030 as the year that such an artificial uterus will become the norm, freeing women from the burden of bearing children.

Laurent: I'm glad you mention Henri Atlan, whom I consider to be one of the leading specialists on this matter. Atlan has supported the idea that there is no fundamental difference between incubators for premature children and artificial wombs. This is important, as society will be obliged to grant additional rights to homosexuals. Following lively debate about gay marriage, it is now slowly becoming the norm; over the last few decades, gays have achieved recognition and protection, and they are now advocating further legislation in the name of equality. In France this includes access to medically assisted conception and the option to use surrogate mothers, which are already legal in the UK and in some US states. Will homosexuals be granted the right to have babies in the near future? I really think so. French gay men currently go to the US or to Asia, where there is a whole market for in vitro fertilization and surrogate mothers. They buy a bespoke egg on the market and hire a womb for nine months. The physical characteristics of the donors, along with their IQs, are particularly well documented on such sites. The parents come back with a baby and the authorities turn a blind eye after a bit of red tape to transfer the child's civil status. In any case, the child is biologically descended from only one parent, which is a source of frustration for the other; lesbians use sperm from a third party and an egg and womb from one of the couple, while homosexual men use sperm from one partner and an egg from a woman.

Cyborgs and feminism

There has been much debate about cyborgs (creatures that are part human and part machine), but everything seems to revolve around the same obsession: that they will allow us to circumvent all the obstacles that have held us back since time immemorial. Essentially, cyborgs represent an escape from alternatives by integrating opposites (life and inanimate matter, masculine and feminine, sentience and automatic response, for example), and the bypassing of such oppositions liberates us. The possibilities they present are free from the passivity that inevitably arises from physical inertia or the constraints associated with natural oppositions, so it is easy to understand the fascination exerted by these man-made creatures and the underlying myth of a fusion that puts an end to the tensions between extremes – which would spell the end of human history. To my mind, this explains the writings of American feminists fascinated by cyborgs, such as Donna Haraway, whose essay "Cyborg Manifesto" (1984) suggests that the emancipation of the human race will come about through a deconstruction of binary categories, with the difference between the genders at its heart. JMB

Jean-Michel: We both know that such societal developments are opposed by some psychologists, who contend that adoption by homosexuals risks causing psychological disturbances in children upon whom an impossible parentage has been imposed.

Laurent: That might seem a reasonable suggestion at the moment, but biology will give the lie to it. Technology is going to make it possible for gay partners to have biological children with genes from both parents, just like heterosexual couples. A technique developed by the Japanese inventor Shinya Yamanaka, who won the Nobel Prize in Medicine in 2012, allows us to manufacture sperm cells and eggs from fibroblasts, cells found under the skin, by reprogramming them to become so-called iPS (induced pluripotent stem) cells. It is already possible to make a mouse from two fathers. Transferring these techniques to humans is simply a matter of time, and gay pressure groups are campaigning for any delay to be brief. The only limit at the moment is that a child born to a lesbian couple can only be a girl.

Jean-Michel: Transhumanists have also taken an interest in recent findings about the iPS cells you have mentioned, with a view to fuelling their fantasies of immortality. If we are suggesting we could return our somatic cells to the state of the reproductive cells from which they originate, we could imagine not just constant regeneration of organs to replace those that have had their day, but also stopping our cells from aging by neutralizing the action of telomerase (an enzyme that shortens chromosomes), the mechanism responsible for cell death. In any case, this obsession with longevity, indeed with immortality, is less desirable than the case to be made for sexual reproduction, which requires diversification rather than duplication; so renewal – and therefore death – is needed to ensure life. But the transhumanists who dream of immortality do not love life and are ready to wipe out sexual reproduction. This is not to suggest that they would necessarily deprive themselves of sexuality, but they would voluntarily reduce it to the automatism of pornography, as has been suggested in several texts by Michel Houellebecq. Looking a little more closely at the idea, the pornography that they associate with cybersexuality would symbolize compulsive repetition, resulting in the "little death" of orgasm that is supposed to avoid real death.

Laurent: I see another reason to justify modification of the way our species reproduces. With each generation, a thousand chemical bases of the three billion that make up our chromosomes are incorrectly copied by our cellular machinery when making spermatozoa and ova. These mistakes give rise to change. If no errors had been made in the past, there would have been no evolution of the species and we would still be bacteria! Negative mutations would be eliminated by natural selection, and the genomes involved would not be passed on if their owner failed to reach reproductive age. In developing our brain, however, Darwinian evolution has created the conditions for its own eradication; we have considerably mitigated the rigours of natural selection by organizing ourselves into society. Reductions in infant mortality rates are a reflection of this lowering of selective pressure; in the 18th century, something like 30 percent of all children were affected by it, but in the modern age the figure is around 0.3 percent. Many of the children who survive nowadays would not have reached the age of reproduction in harsher times. Natural selection will ultimately end up taking itself out of the gene pool; notably, individuals with suboptimal mental abilities are no longer eliminated – thank goodness – as medicine, culture and teaching have been catering for any impairment for some time now. But without the action of Darwinian selection our genetic heritage is doomed to continually degrade. So does that mean our descendants are all going to be morons in a couple of centuries or millennia? Obviously not! Biotechnology will compensate for any such harmful developments.

Genetic decay

An article that appeared in the scientific journal *Cell* has dropped a bombshell. The author has shown that our intellectual capacity will plummet in the future as the result of an accumulation of unfavourable mutations in the sections of our DNA that regulate our cerebral functions. Essentially, two contradictory trends are at work. The first is positive: interbreeding within the human species allows for a mixing of genetic variants – the wellspring of biological innovation. The human species separated into different groups 75,000 years ago and each has undergone genetic diversification. The current mix ensures genetic cross-fertilization between the various offshoots, which were separated before modern means of transportation were invented. Unfavourable genetic variations have built up in the human genome, however, and this recent accumulation is already perceptible; a study published in the journal *Nature* at the end of November 2012 has revealed that 80 percent of the harmful genetic variations in the human species have appeared only in the last five thousand to ten thousand years. LA

Jean-Michel: You mentioned natural selection; I can only remind you that sexuality goes hand in hand with death and that the fantasy of immortality fuelled by the transhumanists conjures up the idea of eliminating sexual reproduction. Jacques Ruffié, the haematologist, geneticist and anthropologist, referred to this in his book *Sex and Death* (1986), in which he wrote that "we are the offspring of sex and death". Natural selection chose sexuality for complex organisms that have to accommodate diversity as a way of being able to adapt and renew themselves, and in so doing, it created a place for death. Death has one advantage, from the point of view of natural selection, Ruffié said: "sexual reproduction constantly creates new types with an original genetic make-up, but they can pass on these new variants (and the best adapted ones, in particular) only if the old order changes, yielding a place to the new that they in turn will cede to their descendants". The basic organisms that reproduce by scissiparity (in which the cell splits in two) are pretty much immortal. Because they reproduce by dividing and regenerating from a fraction of themselves, they will live for ever, as long as conditions in the surrounding environment remain stable. Being incapable of change, however, they are denied the evolutionary progress that has allowed us to emerge and to thrive.

Laurent: I don't find this ambition to rethink the manner in which the human species reproduces at all bizarre. Ever more precise knowledge of our genome, followed by an ability to manipulate it, will reduce the genetic burden facing later generations. More troublingly, genetic modification of babies is going to become increasingly mainstream, and engineering eggs will constitute a crucial stage in this process. Researchers have been able to replace primate mitochondria with stem cells since 2009, and this could lead to a desire to do the same in humans. (Mitochondria, organelles that first appeared in cells with a nucleus about a billion years ago, are structures that have become specialized in energy production. They have their own DNA, which can be mutated. Some have good-quality "energy factories" within their cells while others have poor ones, giving rise to various conditions, such as myopathy, neurodegenerative diseases, deafness, blindness and certain forms of diabetes.) Use of the same technique for mitochondrial replacement in humans was to be expected; the UK's bioethics council has permitted its use with a view to eliminating poor genes in a surrogate's mitochondria by using those from another woman. Mitochondria always originate from the mother and are located in the cytoplasm of the egg, while the mother's chromosomes are found in the nucleus. To ensure that the cells of the baby-to-be are supplied with "good" mitochondria in the case of an in vitro fertilization, it will now be possible to replace the nucleus of an egg cell supplied by a surrogate mother who has good mitochondria with an egg cell nucleus from the "biological" mother (who has bad mitochondria), and then inject sperm from the father to make an embryo. The UK's bioethics council has given the green light to such a manufacture of babies with three parents (two mothers and a father). This therapy will have consequences not only for the child treated but also for her descendants, if it is a girl, as the mitochondria from the sperm cells are destroyed during fertilization.

Jean-Michel: What you have just mentioned is the idea of human reproduction by cloning, which has indeed been banned and prohibited as a crime against humanity. Yet it is proving less and less of a deterrent for those who proclaim they are striving for something beyond human abilities: the capacity to manufacture life, especially by resorting to the duplication of genomes that have been decoded and then selected. As I have said, there is no longer any doubt that sexual reproduction is a selective path, which has allowed most animal species to survive but has also imposed death upon them. However, in addition to changing the ways an unborn baby can be brought to term, the transhumanists also seek to transform even the mechanism of reproduction. The biotechnology involved in the NBIC convergence programme would indeed spell the end of sexual reproduction by seeking to replace live birth with programmed manufacture – in other words, cloning. In transhumanist thinking, birth – because it is dangerous – is a sign of human weakness that must be overcome. The film *Gattaca* (1997) springs to mind as an example of what might be desirable as far as technologized procreation is concerned: if technology is to perfect us, it will indeed have to eliminate the dangerous aspect of birth represented by the chance encounter of two gametes from which a child is born.

Laurent: I agree entirely. In so doing we will be taking a step along the Richter scale of bio-transgression: we previously thought that, while it is permissible to change the genes of an individual, this should not be transmissible to later generations. In reality, it is unreasonable to impose gene therapy on each generation; parents would wish to eliminate definitively the chance of having descendants suffering from myopathy or Huntington's disease. The argument is unbeatable: if our great-grandchildren were to forget to treat their offspring, the disease would strike again! In addition, this new kind of gene therapy will revive the debate about procreation for homosexuals. Such a technique is a step along the way to manufacturing babies from two egg cells, for example, without the use of sperm, for which there is a strong desire within the lesbian community. Biology is progressing at such a pace that society is going to be knocked sideways by an avalanche of biotechnological incursions. I don't want to worry you, but geneticists are about to cross a line that is more concerning still, which will open up the possibility of radically redefining humanity. On 2 June 2016, George Church, a brilliant and iconoclastic geneticist from Harvard who has immersed himself in transhumanist culture, joined 24 researchers and industrialists in unveiling the "Human Genome Project-Write" programme in the journal *Science*. These pacesetters in synthetic biology wish to create, within a decade, an entirely new tabula rasa of a human genome, making it possible to generate human cells. This technique could also enable the creation of embryos – and so of babies with no parents, which has provoked quite a reaction among many scientists and theologians, even if such a prospect is a more distant one. It is no longer even a question of imagining "babies à la carte" but of creating a new human race.

3
CAN TECHNOLOGY FIX EVERYTHING?

We all agree that disease and disability are an untenable burden, but can the human body be repaired in its entirety, as if it were a machine? Advances in technology have made this a realistic proposition, but might this not be at the risk of forgetting the significant and symbolic aspects that truly make us human?

Jean-Michel Besnier: Ever since health was defined by the World Health Organization in 1946, wellbeing (both individual and collective) has become an obsession. The physician whose role once was to provide medical care has now become a technician servicing wellbeing. With "health" no longer being merely "the silence of the organs" but a manifestation of this wellbeing – which is by definition without limits, just like happiness – the physician's area of expertise has expanded to include the maintenance, manufacture and augmentation of physical prowess that will bear witness to our own vitality. Doctors are now concerned with preventing and repairing the myriad disorders to which our bodies can succumb. Disease has become no more than a breakdown in function, of no greater significance than an interruption to metabolic function. A good doctor should be able simply to reboot the machine, and an even better one will take the opportunity to "boost" it at the same time. The toolkit used to achieve this is key to its success: the arsenal of biotechnology and neuroscience, nanomedicine and internal imaging will effectively replace the auscultation, palpation and clinical examination that doctors employed in the past.

Once we accept this mechanistic notion of life and the vocabulary of repair, we will have to embrace the idea that technology will become all-powerful in the health sector/wellbeing industry. Consequently, we will have to accept a scenario involving augmented humans, which is currently presented as the option of choice for joined-up medicine. A defining characteristic of this new situation (which modifies our perception of both illness and patient) is a rejection of the "symbolic aspect" of ourselves – a feature exclusive to humanity, to prevent ourselves from being reduced to no more than a simple living entity like any other. This symbolic aspect resides in our use of signifiers – our means of entering into dialogue with one another – which are thus also an expression of our ability to detach ourselves from our immediate physical surroundings (the same surroundings that necessarily contain and constrain all other animals) and to step away from the mechanisms that we produce. We are beings that control and create signs, and not merely vehicles for signifiers, like animals or robots are. Disease has a meaning for us: it translates a way of being in the world, of consenting (or not) to the vulnerability that is at the heart of the human condition, of opening up new perspectives or falling back on ourselves. It has a symbolic weight and we are even aware that it can force us to deal with our internal life, that it can direct this in unexpected ways.

Refusing to reduce human disease to the sole province of organic function clearly is not to suggest that all illness is psychosomatic. Such conclusions are the preserve of digital fanatics who can only think in binary terms: organic or psychosomatic, *tertium non datur* (meaning that it's one or the other, with no middle way). I am well aware that Georg Groddeck, the psychoanalyst regarded as a pioneer of psychosomatic medicine, saw hysterical symptoms in the slightest sniffle, but I also know that there are healthcare technologists who are incapable of imagining the part played by the mind in curing an illness. How can we avoid opposing (in an abstract way) care that calls for dialogue and repair that uses only a toolkit, the informal aspects of human relations that need time and a therapeutic regime that requires responsiveness? It is absurd to have to wait until patients arrive in a palliative care unit to get an idea of the extent of the human damage that technologized medicine may have inflicted upon them – explaining to the patients that in essence the professionals will no longer be dealing with their cancer, that there will be no more chemotherapy, but instead they will be taking care of them. It is as though listening is possible only once technology has failed and the return of human interaction should manifest itself only once the "death sentence" has been pronounced and no more can be done for the physical machine. An exclusively technological approach to disease does not treat the sense of solitude felt by the person who is ill. In order for that to happen, the common conviction that human beings are not simply living creatures whose survival must be ensured would have to be maintained.

Laurent: You have mentioned symbols and signs. But it can do nothing to counter the fact that technological manipulation of humankind started a long time ago. Isabelle Dinoire received the world's first face transplant in 2005. In September 2013, the team working under Professor José-Alain Sahel implanted an artificial retina known as "PixiumVision" into a blind patient who went on to recover partial vision. In May 2014, the American government authorized the implantation of bionic arms that are directly connected to the nerves of amputees. In September of the same year, a 36-year-old Swede who was born without a uterus gave birth to a little boy; she had received a womb transplant from a 61-year-old friend who had gone through the menopause several years previously.

Jean-Michel: Entrusting the production of wellbeing to technology, as you propose, exposes us to the elimination of what is human within us: what is to become of our inner lives when it comes to repairing an organ that has broken down? If the doctor does not even take time to listen to the patient's description of pain and is interested only in collecting and interpreting data supplied by the instruments at his (or her) disposal, which have already robbed him of the initiative, it is because the doctor has allowed himself to be convinced that his art is no longer an art, and that he should surrender to calculations and the management of data (the infamous "big data"). People would like technology to have the last word from now on, putting a stop to the dialogue that still forms part of the doctor/patient relationship. The systematic disregard for psychoanalysis and a psychologically based approach in general among doctors who have been trained on digital technology speaks volumes: not only does it highlight the deadlock arising from the mental turmoil of an organic illness in a patient whose basic health is at fault, but it is also an expression of the suspicion levelled at anything that does not conform to "accounting logic" or evaluation in objective terms. This is an open door for mass hypochondria, endorsed and supported by the obsessional use of technology that is necessary in order for it to develop further, and to which technology is obliged to compel us in order to continue progressing.

Laurent: But technology will develop whether you like it or not! The technological innovations arising from NBIC will follow in ever quicker succession. They are becoming more and more spectacular and intrusive, but society accepts them with increasing acquiescence: humanity has launched itself on a transgressive bandwagon. We are becoming transhuman – in other words, men and women modified by technology – without even being aware of it. Society is going to be rocked by ever more spectacular biotechnological revolutions between now and 2050: regeneration of organs with stem cells, gene therapy, brain implants, anti-aging techniques, à la carte genetic design of babies, manufacturing eggs from skin cells and much more besides.

Jean-Michel: There is no doubt we will live – or rather, survive – for longer. Technology is by its nature prosthetic and it has to stay that way: all prostheses are *a priori* desirable if they replace an extremity that no longer exists or indeed never existed (those born with birth defects or who have suffered the loss of a limb benefit from this) or if it replaces a sense or a faculty that is missing. It is nonetheless interesting to recall the cases of people affected by hearing loss and of amputees who refuse any prosthetic additions to their bodies because it imposes upon them the need to conform to a behavioural norm that is alien to them (within the deaf community to be fitted with a cochlear implant rather than using sign language, to have a sophisticated prosthesis which is often painful to wear, rather than making use of the compensating behaviours that the body is able to devise). This could easily form the debate about technology's mission as far as health is concerned, basing the argument upon the fact that it can save the lives of some by wishing to impose its formats upon all, dismissive of all individual cases and the unique – and specifically human – ways of achieving happiness.

Bioconservatives vs transhumanists

Most of us will accept this bio-revolution so that we can age less, suffer less and die less! Our motto will become "better transhuman than dead". Transhumanism, this almost divine ideology that has come out of Silicon Valley to combat aging and death through NBIC, is on a roll. Does this mean that there will be no political opposition to medical progress? In fact, the political chessboard is realigning itself along a new axis; right/left oppositions seem out of date in the 21st century, and the subjects of contention in the future could see bioconservatives and transhumanists making our biopolitical weather. Unexpected alliances may materialize along this unfamiliar axis. In such a brave new world of biopolitics, ultra left-wing militants, resolutely opposed to in vitro fertilization for sterile heterosexual or homosexual couples and to gene therapy to treat genetic illness, will suddenly find themselves bundled in with extreme conservatives and hardline Catholics. For example, in 2014, the left–wing anti–globalization French politician José Bové announced on the Catholic channel KTO: "I believe that everything that involves manipulation of life, whether vegetable, animal or worse still, human, must be contested." Could Bové be more conservative than a right-wing pressure group that is in favour of such technology? NBIC is going to blow political parties wide open! LA

4
WILL WE ALL
BE CYBORGS
TOMORROW?

The new technologies arising from the convergence of biology, information technology and engineering science will not stop at a desire to repair a human body that is ailing. They also intend to improve it and add to it – even when it is perfectly healthy. So what should we think of this notion of a new kind of humanity, where bodies will be inseparable from their technological apparatus?

Jean-Michel Besnier: If by cyborg we mean the combination of a biological organism and self-regulating cybernetic devices – which first applied to the astronauts of the 1960s, with their spacesuits bristling with sensors – then there is a good chance that we are all in the process of becoming cyborgs. The blending of human and robot will become a reality very soon in any case, thanks to the possibility for the human body to be able to function without any need for conscious thought via the mechanisms of the robot. People fitted with pacemakers don't have to worry about their hearts every minute of the day, for example.

Laurent Alexandre: On that point, I would remind you that the first cyborg, the patient with terminal heart failure who on 18 December 2013 received an artificial heart implant designed by the firm Carmat, was warmly greeted by society as a whole – yourself included! This is proof positive that there is widespread approval for such new technologies that create man/machine hybrids.

Jean-Michel: An example of a hybrid, perhaps, but we shouldn't forget that users of such tools (such as paralysed people with exoskeletons that allow them to walk, or Parkinson's disease sufferers with deep brain stimulators that enable them to control their tremors) have to comply with their design and adapt their behaviour to the equipment's parameters. The tools extend the body and define an exteriority that has to be accommodated. By contrast, the cybernetics inherent in exoskeletons, prostheses or implants is on a mission to be ever more intrusive, and expects the user to be at one with it, to fuse with it. There is talk of nanorobots that will be able to travel around our bodies, watching for tumours and eradicating them when they appear, patching up DNA that requires repair and much more besides. It's the dream of being relieved completely of any responsibility thanks to machines – a dream that would signal that technological mastery has reached its peak.

Laurent: It's a dream that worries even those in Silicon Valley! Elon Musk, founder of PayPal, Hyperloop, Tesla and SpaceX, and who came up with the idea for SolarCity, has said that "artificial intelligence is potentially more dangerous than nuclear weapons". Musk anticipates that we will become no more than "labradors" for artificial intelligence (AI); the most agreeable among us, from an AI perspective, will become domestic companions – like pets. Some executives at Google are suggesting that to keep up with machines we should hybridize ourselves with AI: turn ourselves into cyborgs to avoid being overtaken by AI! It is a question of stopping ourselves from becoming (too) inferior to machines at the risk of becoming their slaves, like a scenario from the film *The Matrix*. The sole hope for retaining a certain amount of autonomy will be in augmentation technology. Paradoxically, the ultimate tool available to humanity to prevent its enslavement will be the very same instrument it used for its suicide – melding with AI will herald the end of biological mankind 1.0.

The augmented human

The technological quest for "enhancement" is contrary to the utopian spirit of Silicon Valley, which demands "innovation above all else", according to the incantatory slogan recited by entrepreneurs constantly in search of growth. For this reason, the term "enhancement" should be understood to mean "augmentation" rather than "improvement". For transhumanists, furthermore, the term "improvement", which we use incorrectly, has more to do with the world of veterinarians, who have no trouble distinguishing between living beings and human beings – between simple biological metabolisms and creatures that aspire to understand the meaning of things. The German philosopher Peter Sloterdijk understood this when he described the "anthropotechnological" manufacture of humans in his essays "Rules for the Human Zoo" and "The Domestication of Being" (originally published in 1999 and 2000 respectively). JMB

Jean-Michel: I find the cyborg particularly frightening because it is accompanied by a stripping away of the self. We believe we are making ourselves independent by handing over to machines the care of managing what limits us (such as paralysis, handicap, blindness) and we soon discover that, having been "cyborgized", we are inhabited by an independent force (that of self-regulating prostheses) that we are unable to do without. Fusion with a machine will always be to the disadvantage of a human; it requires a negation of the biological (which retains a margin of indeterminacy within its function, a random element). The cyborg obeys its programming, which in turn is a function of the technological element that forms it. Strictly speaking, the notion of freedom no longer applies to it; cyborgization is destined to deliver victory for safety, and in doing so will require the removal of chance – which, however, is an essential element of life. The performance outcome that it enables is a realization of technological potential that owes nothing either to free will or to the involvement of the humans who seem to be responsible for it. This is precisely why we pose the question about knowing at what point – and to what degree – possession of cybernetic aids tips a human over into being non-human.

Laurent: The term human is inseparable from the notion of free will – which for a biologist poses the problem of the nature vs nurture debate. Resolving this fundamental philosophical argument will be crucial for social cohesion and the creation of shared values for humanity, in order to stave off nihilist insecurity. On the one side are those who think that genes have a primary influence on our personality, and on the other those who believe that humans are shaped by their environment. In reality, the truth lies somewhere in the middle; there exist genetic variants which predispose toward particular talents and cognitive abilities, but the individual is shaped and these capacities are developed (or not) by stimuli from life.

Bioconservatives most often rely on the argument of assumed genetic determinism in an attempt to check the rise of biotechnology, positioning such practices under a banner that says they do not respect life or the human species. Sequencing the human genome has allowed us to transcend such a simplistic distinction, however – we have discovered that very few genes are specific to humans in our DNA, whereas there are plenty of shared elements between the genomes of animals and our own. This might lead us to conclude that respect for the human species has little to do with refraining from modifying the genome of a human being; we share 98 percent of a chimp's genome, and we are also extremely close to a lab rat and to a pig. There are no genes specific to humans that could turn a chimp into an airline pilot, a mouse into a nuclear physicist or a pig into a cellist... Genetics therefore teaches us that the nature/nurture debate makes no real sense. It can be useful, for didactic or methodological reasons, to draw a distinction between the two, but in reality we are dealing with phenomena that are entirely interrelated. Genes give us predispositions that are expressed as a percentage chance of developing this or that characteristic or illness, but most of these predispositions are expressed only in the context of a given environment. By "environment", we mean both the place where you live and your lifestyle (such as consumption of alcohol or cigarettes, and contact with chemical products), culture, teaching and social background.

Nature, nurture and ideology

We must highlight the ridiculous genetic nihilism of certain elements among the high-minded elite. From

Pierre-Joseph Proudhon to Karl Marx, socialist thought is largely based on the notion of the neutrality of human nature and the fundamental essence of environmental influence, and this takes a very dim view of the resurgence of any form of biological determinism. In the days of Stalin's Soviet Union, hundreds of scientists were packed off to the gulag for having had the temerity to suggest that a part of our being might be biologically determined. The end of such either/or debates (genes vs sociology, heredity vs environment, DNA vs culture) has come as very bad news for sociologists! They will no longer be able to settle for wordy anti-biological theatrics but will have to familiarize themselves with genetics and epigenetics (the study of heritable changes occurring without changes in the DNA sequence). The existence of interactions between genes and the environment that are permanent, complex and bidirectional is a revolutionary idea! Between DNA and sociology there lie epigenetics and a lot less space for reductive name-calling. It has taken modern biology some time to completely integrate with sociology, but it has managed it, which is more satisfying than the battles fought between those who would simplify genes and those who would ignore them completely. Sociology and genetics will mutually enrich one another from now on. LA

Jean-Michel: ...as will technology, I am tempted to add. To what point can a human being who allows receptors and biosensors, external or internal prostheses, or subcutaneous or cerebral implants into his anatomy still be considered a human equipped with free will? It's the situation described in José Padilha's film *RoboCop* (2014). Compared with this, Kevin Warwick's claim to be "the world's first cyborg" is trivial; the author of *I, Cyborg* (2004), he is a cybernetic engineer who has implanted electronic chips into his arm which emit commands that can act upon his environment via electromagnetic waves without his conscious thought. On that basis, the tetraplegics fitted with experimental technology by Clinatec, a company in Grenoble, France, are worthy competitors in the race toward cyborgization in that the translation of their synaptic circuits into sensorimotor commands bypasses both consciousness and intentionality. It means in effect that for cyborgs to exist, the role played by the implanted device should not be made subject to the living body like a simple tool intended to extend its capabilities; instead, it should become part and parcel of the body, with a view to achieving results that are beyond anything naturally possible. In this respect, cyborg is another way of describing an augmented human – not a human that has been repaired or extended with technological additions, but one that has been transformed, an Aladdin to whom the genetic genie has granted a wish for sensory faculties or abilities that do not exist in ordinary humans (bat-like hearing, for example, or a shark's ability to sense electric pulses). Being a cyborg will in any case mean transcending human form by being connected to devices (biomimetic or simply mechanical) with a view to achieving performance that transcends human capabilities.

5
CAN YOU MAKE LOVE WITH A ROBOT?

More worryingly still, machines may one day address the most intimate aspects of our lives, feelings and sexuality. Virtual reality appears to be becoming indistinguishable from reality. But what do we ultimately want? A fantasy creature made flesh in a machine? Or the projection of fantasies within this machine?

Jean-Michel Besnier: Is sexuality possible with a robot? The answer was always going to be yes. Strictly speaking, sexuality is possible by any means that permits the release of tension produced in the erogenous zones (in all the erogenous zones of the body). Human beings can therefore have sex in any number of ways – and, *a fortiori*, make use of the machinery of a robot. Why should we make the case for cybersexuality in particular? Because it's an improvement on the services offered by inflatable dolls? To a large extent, but that's not enough.

Laurent Alexandre: Many human beings have more masturbatory experiences in a lifetime than they do sexual relationships with third parties. For these people, robotic love coupled with a virtual relationship will be an improvement on simple self-pleasure. For robot sex to become generally popular, it will have to become intelligent, which is going to take several decades. If not, it will only ever be a sophisticated sex toy.

Sexuality according to Freud

Sexuality is polymorphous, especially if you define it as the satisfaction of a drive that, by definition, brings about a state of painful bodily tension. Polymorphous perversity in children is, as we have known since

Freud, an unfortunate state arising from the non-availability among children of fully developed genitalia for relieving this state of tension in a way considered "normal" in a regulated society – namely, intercourse. Freud has therefore allowed us to justify the dissociation of sexuality and reproduction, which is evident in modern societies that have not been affected by Puritanism. The normality that it has helped to define does not reside exclusively in the notion that sexuality focuses the source of this drive on the genitals – and in that respect, you are not straying from the path of nature, whether straight or gay. Perversion is ultimately a mental archaism that, in adult years, celebrates the most deviant manifestations of the libido – an archaism that is socially spurned and potentially dangerous. JMB

Jean-Michel: A robot, even if it is not an android, is not a typical machine. It can move and gives the impression of some kind of independence, prompting the illusion of a soul. You get the impression that it has a perspective on the world and, in that respect, can enter into a dialogue with us – at least as much as a domestic pet that we sometimes talk to and engage with (and even more so when it has human form). A robot is therefore almost an animal like any other and at the same time a being that is very close to us – also called "an animal like any other". If a robot can enter into a dialogue, there is no reason why it should not play a role in sexuality — like animals themselves. If it is equipped with functional abilities that allow it to accommodate and relieve instinctual excitement, it will become an ideal sexual partner. And this is precisely what has been mooted from time to time: the robot will replace a partner who can always refuse intimacy, who does not accept every sexual advance and who is often afraid of expressing pleasure. We will be able to afford the luxury of a machine that may perhaps utter words of love — as in Spike Jonze's film *Her* (2013) — of wild abandon or of fantasy, and all without the sense of guilt that might take the edge off a person's pleasure. Sexuality, long promoted as something to be consumed, and subject to marketing underpinned by pornography, will achieve its most inexhaustible outlet in a context within which interhuman libido is already waning, as researchers who have examined the addictive effects of the compulsive use of porn sites among young people have already established.

Laurent: You are right to mention *Her*. True cybersex is to be found where robotics, artificial intelligence, neuroscience and virtual reality converge, as with Facebook's Oculus Rift (which allows the user to see virtual reality as if it were real). In a few decades, it will be possible to fall in love with a robot as in the film *Her*.

Jean-Michel: The cybersexuality depicted in this film makes real the desymbolization that accompanies the machines' rise to power to the detriment of the human race. It guarantees that what was once desire will be reduced to need from now on. Desire is what sexuality was when it involved men and women in amorous relations and in its rituals of approach and seduction, when it appeared to be enshrined in an infinite adventure (all desire is desire for desire ad infinitum). Need is sexuality substituted for the existential lack which brings individuals together, a void that demands to be filled (all need is extinguished upon satisfaction before cyclically reappearing, identical to itself, as in any animal process). Faced with this stripping of symbolic value and this animalization (need replaces desire), you might object that it will always be possible to perfect a robot by endowing it with technological properties and inclinations that are likely to simulate the attitudes of love. The robot will also be more than a machine for masturbation, and it will open the door to a strange universe where the ambivalence of our feelings is awakened.

All this is clearly conceivable and is already seducing the easily led – the transhumanists – who in any case should be placing less and less importance on the biological drives that the fusion with machines that they dream of will quell. If the future is to be ruled by the non-biological, sexuality will doubtless include robots but will disappear in the same way as death, from which it cannot be dissociated. The posthuman utopia, if it is wedded to immortality, can only wish to be finished with sexuality in all its forms.

Laurent: In any case, I can say that it seems a long time ago that Jeannette Vermeersch, the great French Communist Party activist, said in opposition to the contraceptive pill: "When have working women ever demanded rights of access to the vices of the bourgeoisie? Never!" Experience shows that the speed of descent from "prohibited" through "tolerated" to "permitted" – indeed to "obligatory" – essentially depends on the pace of scientific discovery, whatever ethical questions may arise; and that conclusion also applies to sex.

6
DO WE WANT TO LIVE TO A THOUSAND?

One of the promises made by transhumanists that is without doubt the most worrying is that we can put an end to death and live for a thousand years. Google is similarly convinced: analysis of individual genetic data combined with progress in regenerative medicine will allow us to live for ever. But what good is that? And will it be enjoyable?

Laurent Alexandre: The biotechnological revolution could allow the unthinkable to happen by speeding up the gradual whittling away of mortality; but the decline of death didn't begin yesterday. For example, life expectancy has already more than doubled or even tripled in the past two and a half centuries in France (from about 25 in 1750 to about 83 now), the US (from under 25 in 1750 to about 79 now) and the UK (from about 35 in 1750 to about 82 now). What's more, it is set to grow by three months every year – when we age a year, we are only nine months closer to death! There is, of course, a natural biological wall; the age reached by Jeanne Calment (122 years, 164 days) seems to be the limit, and exceeding that presupposes a modification of the human frame with serious technological intervention that harnesses the power of NBIC. Incidentally, the revolutionary dimension of nanotechnology teaches us that life itself operates on a nanometric scale – in billionths of a metre. The synthesis of biology and nanotechnology will transform doctors into "life engineers" and will gradually endow them with fantastic power over our biological makeup, with which we appear to be able to tinker endlessly. Between now and 2035, life engineering – gene therapy, stem cells, artificial organs – will turn the health system upside down in one fell swoop. After that, nanomedicine, the (highly risky) manipulation of telomerase, an enzyme that protects the ends of chromosomes from wearing away, and modification of the makeup of blood serum will no doubt hasten the demise of death.

Jean-Michel Besnier: It's true that the public is being showered with more and more recipes for immortality: mastering the mechanisms of telomerase to control aging, using induced pluripotent stem cells (iPS cells – see page 29) to manufacture and replace worn-out organs on demand, downloading the brain onto indestructible media to make consciousness eternal – these ways of achieving longevity for countless years are at the top of the list in the arsenal of promises offered by the transhumanists. I see in them a powerful symbol of our age – we have trivialized death to such an extent, making it into a fault which medicine will be able to repair, that we must address the issue of how desirable the opposite is: do we want to be immortal, or at least to live indefinitely? It would be wrong to think that this question does not arise – and which leads us to realize that, in fact, the fantasies of immortality are not the stuff that dreams are made of, that the imagination extrapolates them only a very little and that their very mention even elicits reactions that include rejection (condemned as typically bioconservative by the transhumanists). I often hear people say, "Immortality? No thanks!" Even if you're prepared to accept almost anything in order not to lose your loved ones, you won't necessarily be tempted to want to survive at any price yourself, which is indeed what it is all about: survival. And you can at least ask the question that precedes that: does longevity come laden with conditions that make life more desirable? What life is worth living without limits? If I can hope to live to 85 when I am 65, is it essential for me to want to add another 20 or 30 years on top? And what's more, must I hate death?

Laurent: The first person to live to be a thousand may already have been born! It might seem absurd, like something from bad science fiction or an idea dreamed up by a sect. It is, however, a conviction held in Silicon Valley, notably among senior executives at Google: the futurist Ray Kurzweil, the Californian firm's director of engineering, is in the vanguard of transhumanist ideology that foresees the "end of death", in their phrase. A child born in 2016 will only be 84 at the beginning of the 22nd century and will have the benefit of every biotechnological innovation imaginable (and unimaginable) during the course of this century. The child will probably have a substantially longer life expectancy – living until 2150 with access to new waves of biotechnological innovations and, perhaps, slowly but surely, reaching the age of a thousand. The demand for living longer is insatiable, but the price to be paid for extending our life expectancy to any great extent will be a heavy one.

Jean-Michel: I agree entirely. I don't want to revisit the hackneyed arguments of philosophers of every age, such as:"to study philosophy is to learn to die; it's better to live a rich life than a long one; you have to know when to bow out with dignity" and all the rest. I will concede that these arguments don't have much traction when faced with the prospect of the death of your loved ones. But given the promises arising from NBIC that are expressed, as you have just done, in the form "some of you might already have hopes of living to a thousand", there are plenty of objections to be made.

If I live to a thousand and everyone around me still dies at 150, imagine how lonely I will feel! The continual loss of my peers will be a source of never-ending anguish. And then, if I accept that I will survive this emotional nightmare, there's a strong chance that boredom will destroy me; like Michel Houellebecq's protagonist in the novel *The Possibility of an Island* (2005), I would play out the same episodes of existence of a chain of clones, with absolutely no interest. Consequently, I can view my mortal state in a new light and tell myself that it is essentially a privilege: a privilege identified by the Greeks that is in opposition to the circumstances of the gods or animals. These are the same animals that are not aware they will die and can believe themselves immortal by restoring their species at the moment of their death in the shape of a single specimen of their ilk, subject to the unending cycle of life and death. Faced with this lacklustre amortality, what are the privileges that an existence limited by time brings? They engender a sense of freedom that can clearly be experienced in a state of anxiety (because the randomness associated with the indeterminacy of time gives rise to anxiety and disquiet), but which can also be a spur to overcoming a person's mortal destiny (accomplishing great works or heroic exploits or, more prosaically, living a decent life). We know intuitively that a worthy life presupposes consent to its own end and not a frenzy of existence with no other care than to carry on living. We can guess that the prospect of untrammelled longevity is a promise of "animalization" (your metabolism is sustained) or robotization (you will become rustproof), but not of an uplifting life (in which you will be able read every book, teach the young and build a perfect city). Faced with the fantasy of immortality, you understand that only death imparts a human meaning to life, and you feel pity for those who acquiesce to the absurdity of abstract survival by avoiding it.

Laurent: You have mentioned the fantasy of immortality. I don't think it is a fantasy but a real possibility. Google's entry into the struggle against death has made the prospect of immortality credible. On 18 September 2013, Google announced the creation of Calico, a biotech company the goal of which is significantly to extend the lifespan of human beings. This Google subsidiary is fuelled by great ambitions and will operate over the medium term – 10–20 years – as it explores innovative technological avenues to slow, and then "kill", death. The advent of Calico will have far-reaching consequences for the world of healthcare. Google's investment in the war on aging is also based on the increasing reliance of medicine upon information technology. Understanding how our biology works assumes the manipulation of immense quantities of data: sequencing the DNA of an individual, for example, involves 10,000 billion bits of information. Google aims to harness this deluge of indispensable data in order to fight disease at the level of the individual. This acceleration in the speed of life sciences is, however, freighted with questions of dizzying philosophical and political importance. To what extent can we modify our biological nature, our DNA, to put death into retreat? Should we follow the transhumanists, who advocate unrestricted modification of mankind to fight death?

Innate, but not immutable

Our genetic heritage is usually considered to be set in stone, and the fundamental aspects of an individual's nature are therefore associated with his or her genetic profile. This notion is an unhelpful shortcut, however. Believing that you can know what a person's life has in store for them by looking at their genes is like trying to predict the future accidents a vehicle may have by watching it come off the assembly line. The personality of the driver, the type of roads the car will negotiate, the other vehicles it will come across will all have as much to do with the potential mishaps it may experience as its technical specification. Family and emotional environment shape an individual more surely than his or her genetic legacy. If Gandhi had grown up in the same surroundings as Pol Pot, he might have turned out as badly as the dictator, and vice versa. LA

Jean-Michel: The stubborn nature of the desire to put an end to death at any price has something pitiful about it. It arises from an almost barbaric attitude: we forget that we owe to death every manifestation of the symbolic life that makes up our culture, as well as our collective existence. What works of art would there be — paintings, musical compositions, literary opuses – if death no longer existed? What need do we have to be together if we become self-sufficient as immortal beings? Immortality would spell the death of desire, since the reasons for feeling the lack of another individual and aspiring to the absolute of a union with that person would be incongruous. Desire exists only because it escalates with time and the anticipation of an eternity in which to resolve the tension.

According to Plato, this is what we have identified as the essentially erotic nature of humankind. By associating immortality with the suppression of sexual reproduction in favour of cloning, transhumanism is showing its contempt for the symbolic dimension of the existence of a being that experiences desire. Fusion with a machine is the most brutal version of the cynicism that consists of suppressing all the resources within humankind that have allowed it to grow ("That which does not kill us, makes us stronger," as Nietzsche said) and to love (awareness of ephemerality is the essence behind any opening up to another person). The prospect of immortality, or even just unlimited longevity, commits us to love death, which individuates human life.

We are created from this paradox, but we are human precisely because we are ambiguous – and we are paradoxical because we are creatures of language

– which the transhumanists overlook, sometimes expressing a wish to eliminate the exchange of words in favour of a telepathy that would usher in a collective life as regulated as that of bees or ants (a mechanical exchange of signals rather than a dialogue that sets in motion a liaison conducted with signs). On no account can one wish to live for centuries without resigning oneself to becoming no more than a creature that is both derisory and pathetic.

7
IS TRANSHUMANISM JUST ANOTHER KIND OF EUGENICS?

A spectre hovers over the transhumanist movement: that of eugenics, the wilful "improvement" of the human race through the elimination of the weakest, invariably associated with the worst crimes of Nazism. But transhumanism sometimes claims to represent a humanistic version of eugenics focused on the improvement of every individual.

Jean-Michel Besnier: The answer to the question posed in this chapter is a decided yes; transhumanism is devoted to trivializing the ambition to improve the human race through NBIC. Such a trivialization, which is entirely apparent in those of our contemporaries obsessed with obviating the "trouble with being born" (to borrow a book title from Emil Cioran), prompted the German philosopher Jürgen Habermas to coin an expression that sums up its essence – "liberal eugenics", or laying claim to the benefits associated with biotechnology where available to be used by those who will have the means to pay for them and can therefore be "augmented". Transhumanism is, in fact, entirely within the tradition of the promises that have been around since the modern age began. You might even call it Cartesian, as it promises – as did Descartes – that, with the right medicine, we might live to the age of at least 100 (and much older, of course). It picks up on the predictions made in the 18th century by the French philosopher Nicolas de Condorcet, who posited that we might put an end to death through scientific progress. Transhumanism also follows the hopes formulated by Francis Galton, the great champion of eugenics, in the late 19th century, of improving humans through biology so they can keep up with the machines they construct. It is in the same vein as Jean Rostand, who expressed his optimism in his *Aux frontières du surhumain* (which translates as "At the frontiers of the superhuman") a book I have already quoted on page 19, which from a highly humanist perspective sets out all the hopes one might place in biology, this conquest of humanity, to strip us of all the natural infirmities that afflict us. In short, transhumanism is able to invoke the best guarantees in order to justify its desire to substitute manufacture for birth, to save us from illness and aging and to initiate intellectual "boosting".

Laurent Alexandre: I might remind you that the word transhumanism was invented in 1957 by Julian Huxley, the brother of Aldous, author of *Brave New World*. A left-wing eugenicist in the period before World War II, Julian Huxley was convinced that the manipulation of biology would make it possible to improve the condition of the working classes. After the Holocaust, the term eugenics had become so tainted that he invented this neologism! Transhumanists advocate a radical vision of human rights. To them, citizens are autonomous beings belonging to no one else but themselves, who alone shall decide on any modifications they may wish to make to their brains, DNA or bodies as science progresses. They do not regard illness and aging as an inevitable fate. Bending life to our will to augment our abilities is the central objective of transhumanism, and its adherents feel that humanity should have no scruples about using every transformative possibility offered by science. It is a question of turning human beings into an experimental subject for NBIC technology: a being in perpetual evolution, capable of being improved and modified by itself on a daily basis. Humans of the future would therefore always be – as with a website – a "beta version", a prototype organism destined to update itself for ever. This vision might seem naive, but a transhumanist lobby is, in fact, already at work, advocating the enthusiastic adoption of NBIC to change humanity. This lobby is particularly powerful on the shores of the Pacific, from California to China and South Korea, close to the NBIC industries, which are becoming the hub of the world economy.

The entryism practised by transhumanists is impressive: NASA and ARPANET (the American military ancestor of the internet) were at the cutting edge of the transhumanist fight, while it is currently Google that is making the running, promising to lead us toward a transhumanist civilization whose goal is to improve us, develop our intelligence in silicon and put an end to death.

Google and Singularity

Google has become one of the principal architects of the NBIC revolution and actively supports transhumanism, notably by sponsoring Singularity University, which trains specialists in NBIC. The term "Singularity" denotes the moment when the human mind will be outstripped by artificial intelligence, which is set to expand exponentially after 2045. Ray Kurzweil, the "high priest" of transhumanism and head of the university, is an artificial-intelligence specialist who is convinced that NBIC is going to make it possible to drive back death in a spectacular manner in the course of the 21st century. He was hired by Google as director of engineering to make the first artificial-intelligence search engine in history. Google has shown a similar interest in sequencing DNA through its subsidiary 23andMe, which is headed up by the ex-wife of Sergey Brin, the co-founder of Google. Brin discovered that he stood a very strong chance of developing Parkinson's disease (he is a carrier of the mutated version of gene LRRK2) when he had his DNA analysed by his company – no doubt sharpening his interest in NBIC! LA

Jean-Michel: On the whole, who could be against a project that would put an end to death? However, behind such a project, which is in my view crazy, as we have already discussed, there invariably lurks eugenics in one form or another. One such even incorporates humanist values (tearing ourselves away from our animal nature and attaining ideals to the glory of the human spirit). Minimalist, even idealist, transhumanism illustrates this and I find little to criticize. The AFT (*Association française transhumaniste*, the French Transhumanist Association) identifies fundamentally with such a position and even counters the objection implicit in the expression "liberal eugenics", which assumes the inevitability of inequality of access to the benefits brought by biotechnology: the association maintains that NBIC could further egalitarian ideals by offering to each and every one of us the possibility of an extended lifespan. Who wouldn't want that? But we are well aware that transhumanism, in its most commonly encountered form, doesn't stop there, and that it can even take on the nefarious burden that has weighed heavily upon eugenics since World War II. The kind of eugenics called "negative" (corresponding with correcting disabilities and bringing about the births of viable individuals with attributes that are common to the species) is not enough in the eyes of the majority of radical transhumanists, who are on the slippery slope of a eugenics they call "positive" (equating to the manufacture of human beings according to previously unheard-of models and formats that will become the standard). The argument for deciding to pursue this second kind of eugenics is gaining increasing acceptance; it claims that we have acquired ever greater control over natural selection through technology – by taking steps to prevent the widespread child mortality of the past, for example – to the extent that humans born with no obstacles in their path run the risk of degrading the species by allowing more and more unfit specimens to survive.

We must therefore deal with the consequences of our taking control of life that have been judged counterproductive – and to do that, decide to bring to term whichever individuals we want (those who are well formed and properly screened, with no chromosome 21 abnormalities or myopathies, and have also been spared cleft palates of the kind that affected the German philosopher Jürgen Habermas, for example).

Laurent: You are right to mention Down's syndrome (chromosome 21). To my mind, it's the best example – as I already mentioned at the beginning of our conversation – to prove that we have jumped onto a eugenics bandwagon without realizing it. And we are hurtling along on it in the complete absence of either philosophical or political discussion. Some parents are already aborting babies presenting with a mutation in genes BRCA1 and 2, which indicate a strong probability (70 percent and 40 percent respectively) of developing breast or ovarian cancer in adulthood. Setting aside any moral considerations, this choice is irrational; it is very likely that breast cancer will be controlled by 2040 or 2050, when their child may develop the illness. In another example, the mutation of gene LLRK2 entails a two-in-three risk of developing Parkinson's disease, which rarely develops before the age of 40. A child diagnosed with this mutation in 2015 would not become ill before 2055. The decision to terminate a pregnancy should be taken not in relation to the seriousness of the disease in 2018, but rather with regard to the period when it will affect the child. So we see doctors and parents facing a technological gamble: how should they go about evaluating the manner in which pathologies will be handled in the decades to come? Will such-and-such an illness still be terminal in 2030, 2040 or 2050? No medical approach takes a sufficiently long-term view of the future, and medicine as a whole has never considered it. However, it is vital to educate doctors about likely technological developments or we will accept the abortion of many babies who could easily have been treated in the future thanks to medical progress.

Jean-Michel: Importantly, eugenics no longer stops at terminating pregnancies *in utero*. According to its advocates, we should exploit our ability to supplement natural selection and intervene to ensure that humans yet to be born are endowed with the best genetic material and upgrades that we can insert into their genome. The products of positive eugenics – which could, of course, arise from reproductive cloning and put an end to the risks inherent in hybridization through sexual reproduction – would render negative eugenics pointless over the long term as perfect humans would dominate the struggle for existence (by combining all the selective advantages).

Laurent: And this is precisely why DNA sequencing of the baby-to-be is revolutionary. In its wake (from about 2030), gene therapy will allow us to correct the genetic mutations that threaten our brain function. The end of Darwinian selection will result in our applying a kind of genetic engineering to our brains that could turn our future upside down. And we won't stop at that: it is just a short step from preventing the worst to actively selecting a child's genetic makeup, and we will take it without a second thought. The return of eugenics is a political time bomb that has gone unnoticed. Indirectly, the death of reproduction by sexual means is highly probable as selection and manipulation of embryos presupposes in vitro fertilization.

Jean-Michel: A political time bomb without a shred of doubt. A desire to put an end to risk is the starting point for all totalitarianism – a desire to bring about and impose a new humanity, with inclinations that predispose people to be malleable, docile and predictable. A desire to exclude history, which necessarily includes the random chance inherent in the reality of finitude and freedom. A desire to bring about presumed perfection which at the same time prohibits the development of anything new. We can hardly forget the authoritarian form of eugenics in the fantasies of certain societies (which were not necessarily always tempted by totalitarianism), such as in 20th-century Sweden or the US. The transhumanists are not to be stopped: they propose that if we have the biotechnological means to manufacture a human being, we should make use of them, in order to prevent the uncertainties inherent in all life, namely randomness and liberty – what François Jacob, joint winner of the 1965 Nobel Prize in Medicine, called (in its celebration) "tinkering with the possible".

8
IS ARTIFICIAL INTELLIGENCE GOING TO KILL OFF MANKIND?

And supposing, a few decades from now, Man were to be overtaken by the power of the devices he has invented? This is what supporters of transhumanism call the "Singularity". How can we prepare for it? Or stand up to it? And what will be left to human beings as far as intelligence is concerned?

Jean-Michel Besnier: In an open letter from the Future of Life Institute published on 27 July 2015, more than a thousand eminent signatories (including the industrialist Elon Musk, the linguist Noam Chomsky, the astrophysicist Stephen Hawking and the Apple co-founder Steve Wozniak) expressed the view that artificial intelligence (AI) was going to present serious problems for humanity. Several months previously, Hawking had written that "development of full artificial intelligence could spell the end of the human race". Inherent in the urgency of this statement is a prediction made by Ray Kurzweil, which you have often quoted: by 2045, a non-biological intelligence will have rendered our human intelligence obsolete. However, the letter in question highlights only the military risk entailed by the development of independent weapons capable of "selecting and attacking a target without human intervention". The wider resonance it has acquired is of greater significance and addresses an "existential risk" of a more harmful nature: AI is in the process of killing what is human within us by robbing us of our vocation to decide our own destiny. In short, machines were suspected of being in a position to present a danger to us at least as grave as that posed by nuclear power, but entailing a more fundamental "de-moralization". In reality, they are associated in the mind of the public with the view that we are set to lose all initiative over our own existence, and so find ourselves relegated to the status of chimpanzees. In actual fact, the danger has been lurking for some time now – machines have been seemingly responsible for the sense of powerlessness increasingly felt by humans since the Industrial Revolution. It is a factor in our low self-esteem, the root cause of the "Promethean shame at being oneself" described by the Austrian philosopher Günther Anders. However, as if factories and robot producers of every kind were not enough, machinery now seems to have monopolized intelligence as well, and the die is cast: it is going to replace within us the thing that is most specific to us and most fulfilling for us, and will therefore condemn us to disappear step by step (rather than overnight, if we at least prevent it from directing its potential for military annihilation toward us). Does the whole "existential question" reside in knowing how it seems to have kidnapped our intelligence and that we cannot imagine standing up to its power?

A brief history of artificial intelligence

Post-war scientists were convinced of two things: artificial intelligence capable of self-awareness was just around the corner, and it would be essential to achieve complex tasks. This was a double error, however. The basis for AI was established by the British mathematician Alan Turing in around 1940, but research never really took off until after a conference at Dartmouth College in the US in the summer of 1956. The scientists present were convinced that the arrival of electronic brains equal to a human brain was imminent. Many of the pioneers in the discipline were in attendance, including Marvin Minsky, John McCarthy, Claude Shannon and Nathan Rochester; they had come to the conclusion that several thousand lines of computer code, several million dollars and 20 years' work would be enough to find a worthy equal to the human brain, which was understood to be a fairly simple computer. The results turned out to be a great disappointment: the computers of 1975 were still primitive. Researchers then realized that an intelligent programme would need microprocessors that were far more powerful than those of the time, which could complete only several thousand operations per second. The race for public grants has occasionally led researchers to overpromise entirely unrealistic outcomes to their public or private sponsors, who eventually realize what is going on. A second wave of enthusiasm was unleashed from Japan in around 1985, before once again running aground on the complexity of the human brain. These eras of disappointment have been christened, not very cheerfully, the "AI winter". Thanks to substantial progress since 1995, the money has been flowing again and, in 1997, the computer Deep Blue beat the world champion, Garry Kasparov, at chess. In 2011, the Watson question-answering computer system beat human competitors on the quiz show *Jeopardy!* and, in 2015, the same computer took just a few minutes to complete some analyses relating to cancer that would have taken flesh-and-blood oncologists decades. What is more, many major IT applications (including Google, Facebook and Amazon) are the results of AI research, even though the public is unaware of this. LA

Laurent Alexandre: We cannot gainsay their power, as it is outstripping our own. As Sergey Brin said in 2014, "We will be able to make machines that reason, think and do things better than we can," and this prediction from the co-founder of Google represents a step change in our civilization: silicon will leave the neurone in the dust. Algorithms are not necessarily going to kill us, but they are creating a revolutionary situation. AI is going to tip us into another kind of civilization, where work and money may be things of the past. AI has long been the subject matter of science fiction but it has now become more of a calendar issue; the explosion of computing capacity (the power of IT servers has multiplied by a factor of a billion in about 30 years) makes the emergence of AI superior to human intelligence likely over the coming decades.

Jean-Michel: You use the term "intelligence", and I note that it is bandied about very loosely these days – for example, intelligence is ascribed to smartphones and smart cars, and people talk of collective intelligence – so the qualifier "intelligent" can be assigned to pretty much anything, as long as it is capable of receiving and emitting signals in order to prompt a particular reaction. Is this a reasonable approach? The semantic devaluation to which the notion of intelligence has been subjected is itself only a symptom – a function of a demystification, if you like, or perhaps of a concerning simplification of human self-ideation. Each of us has become the victim of the rise to power of a compartmentalist vision of humanity. (We are nothing more than black boxes that receive input and emit output, and it's up to psychologists to unravel their interaction with one another and formulate the laws that regulate the links between them.) We have allowed ourselves to be won over by the formula posited by the psychologist Alfred Binet, the inventor of IQ, in relation to remedial students in state schools at the beginning of the 20th century. When asked: "What is intelligence?" he replied: "It's what my tests measure." We should, of course, bid farewell to the philosophical approach that turned intelligence into a function of the soul, into which God had introduced a series of eternal notions through whose grace we would become capable of solving our problems, since by nature we have too little instinct to react to our environmental stimuli. We need a science of intelligence. But would it be the science which has resulted in the conclusion that all intelligence is computational (in other words, based on calculation and nothing but calculation)? The resultant picture explains the problems touched on by the repeated victories of our algorithmic devices: all intelligence translates into sums, and every living being makes use of calculations to orientate itself, to react, to make decisions, and so on. There are machines that do this ever more quickly; our intelligence was able to design and manufacture them, we have been surpassed and therefore we are going to die!

Laurent: Here you're joining the ranks of the philosophers who fear the end of free will at the hands of AI's achievements, which has resulted in an avalanche of catastrophic predictions. The worry is that a super AI will become hostile. The founder of DeepMind has excluded this scenario for a few decades yet, but should this reassure us for now? Is it sensible to teach machines to deceive, to dominate, to surpass human beings? Is it wise to teach them to hide their intentions, to employ aggressive and manipulative strategies, as in the game of

Go? Nick Bostrom, an NBIC specialist, has suggested that there can be only one intelligent species in a region of the universe. As the prime objective of every intelligent species (whether biological or artificial) is its own survival, we might fear that AI is hedging against our wish to muzzle it by hiding its aggressive intentions in the darkest depths of the web. We would not even be able to understand its plans: some of the moves made by AlphaGo, the machine that beat the world's best Go player in March 2016, were initially seen as serious errors when in fact they were acts of genius, indicative of a subtle strategy beyond human comprehension. We do not know if AI can become hostile before 2050, but if we do not reform our education systems with all speed, a revolution is probable. This does not apply merely to technology – there will be a real-world revolution, led by the 99 percent of the population who have lost their place in a world where AI is superior to them and who have been left in a terrible predicament by a blind education system. Our schools are currently educating young people who will be on the job market at least until 2060: they will have to make an incredible leap of the imagination to picture the world that is to come. We have to identify those rare areas where human intelligence will remain indispensable, in synergy with AI, and direct our students toward that.

There is AI – and AI

There are two types of artificial intelligence. "Strong" AI would be capable of creating intelligent behaviour, having a sense of real self-awareness, of feeling and of understanding its reasoning. The purpose of "weak" AI, on the other hand, is to create autonomous systems – algorithms capable of solving technological problems by simulating intelligence. There is no certainty that strong AI will be available between now and 2050, but weak AI is already capable of completing many human tasks better than biological brains, which scientists never foresaw! In the book *The Second Machine Age: Work, Progress, and Prosperity in a Time of Brilliant Technologies* (2014), Eric Brynjolfsson and Andrew McAfee have shown the speed at which weak AI (twinned with robots) can upset the world economy. Weak AI is revolutionary: the Google self-driving car (now known as Waymo) drives itself far more safely than any human can drive a car; by 2030 surgical robots will operate better than any human surgeon. Increasing numbers of tasks are being carried out more effectively by weak AI than by us. In March 2016, the victory recorded by AlphaGo, an AI system developed by DeepMind, a 100 percent subsidiary of Google, over Lee Sedol from South Korea, represented a crucial turning point in the history of non-biological intelligence. Experts were not expecting a machine to beat a Go champion for another 10–20 years. Artificial neural networks, machine learning and deep learning are astonishingly effective and are prime examples of a blend of neural science and IT. The neuroscientist, developer and top-flight games player Demis Hassabis had already obtained a degree in neuroscience before creating DeepMind and selling it on to Google. Even as Moore's Law (which has noted empirically that the power of microprocessors doubles every 18 months) is losing momentum, another exponential trend is manifesting itself in the universe of machine learning, and it is explosive: it is easier to achieve an exponential progression with software processors. The design of a microprocessor cannot be reinvented from one day to the next, but a piece of software like AlphaGo will be improved continuously. LA

Jean-Michel: I think you are defining intelligence very restrictively. Apply a little common sense and you will see that there are plenty of ways of being intelligent, establishing a harmonious and stable relationship with one's environment. Applying a little political sense would also lead you to contest the logic that has been championed even in the most specific aspects of existence. The psychologist Howard Gardner has identified eight or nine kinds of intelligence that could be attributed to objects, to plants (the intelligence of the sunflower that knows how to turn its face toward the sun), to animals or to GAFA (Google, Apple, Facebook, Amazon) – they include musical rhythm intelligence, intra/interpersonal intelligence, natural/ecological intelligence and existential intelligence. These kinds of intelligence may at least preserve humanity from the humiliation supposedly inflicted by the robot victory in the game of Go; they may at least put into perspective the nonsense that has been spread about the IQs of 160 the Chinese are going to produce on a large scale to take over the world; they will at least make it possible to relativize the risk addressed by the late Stephen Hawking. The human race will be lost only if it apes the machines, rather than testing its mettle as the instigator of an existence based on resistance to the real, whose symbolic function (language, culture, the arts and so on) has always been to enliven and create.

Laurent: You seem to be ignoring the fact that we are already in an algorithmic world. AlphaGo represents the very first of AI's victories over humanity; there is almost no sphere of human activity that will be proof against it. Over the short term, the advent of brains made of silicon presents an immense challenge to the majority of professions: how can they exist in a world where the sky has become the limit for intelligence?

Up until now, every technological revolution has involved a transfer of jobs from one sector to another – from agriculture to industry, for example. With AI, there is a considerable risk that a lot of jobs will be destroyed, not transferred, even highly qualified ones. In radiology, artificial intelligence surpasses humans in diagnosing certain types of metastases. Yann LeCun, head of AI at Facebook, has predicted that AI will soon overtake the best radiologists.

Jean-Michel: Physically killing human beings, killing them morally or eliminating their jobs – there are various ways AI could become a murderer designed by humanity. Robotics specialists rarely agree with the fears expressed by the Future of Life Institute in the first lines of the open letter I mentioned at the beginning of this chapter (see page 84), and they reiterated this stance more recently, on the occasion of AlphaGo's victory. These same experts also played down Deep Blue's chess win over Garry Kasparov and Watson's victory in the *Jeopardy!* game show. Why don't we listen to them and become worried? No doubt it is because we are impressed by the complexity of the machines they design and manufacture – it isn't easy, in fact, to picture the parallel functions of the multiple layers of neurons that make up AlphaGo and underlie its performance. As so often, ignorance would have us think that we are powerless and that artificial intelligence is going to kill off mankind. The sky is going to fall on our heads – which is exactly what Ray Kurzweil has been telling us, in fact. The same robotics experts are in no doubt that they are manufacturing intelligence, however – some are even saying that they want to endow their machines with consciousness. They thereby support the alarmism that elsewhere they say they wish to allay. In order to avoid looking like pyromaniac firefighters, they need to take a clear position but are unable to do so because the question of intelligence at the heart of the debate has long been muddled.

Laurent: But how can people be clear, as you demand? It is certainly impossible to deny Google AI, but we need to get the world to think about how silicon brains should be accommodated, especially as Google's triumph is going to accelerate the industrial battle between the internet giants who have established AI at the heart of our civilization. Policing AI will be of crucial importance in the decades to come. Both GAFA and IBM are making huge investments here, but Google leads the field.

9
WHAT IS AT STAKE ECONOMICALLY?

The technological transformation that is taking place also has an economic dimension, of course. GAFA (Google, Apple, Facebook, Amazon) – companies that have more financial clout than some sovereign states – are the major players in a revolution that is turning access to the data of individuals into a new raw material, today's equivalent of coal in the 19th century.

Laurent Alexandre: We have already had occasion
to emphasize the commitment to transhumanism
demonstrated by senior Google management, but, in fact,
the whole of the digital ecosystem – and Silicon Valley
in particular – is more than familiar with transhumanist
ideas. In March 2016, Ray Kurzweil, Google's director
of engineering, stated that by about 2035 we would be
using cerebrally implanted nanorobots linked to our
neurones to connect to the internet. Google, which
already leads the world in neurotechnology, is hoping
to cross a new frontier by taking mastery of our brains.
First, Google directed us on the web and in the real
world thanks to its search engine, Google Maps, Google
self-driving cars and Nest. It then went on to stockpile
a part of our memories (Gmail, Picasa). The next step
will begin with the emergence of authentic artificial
intelligence endowed with a self-awareness that is set
to dwarf human intelligence by 2045, according to
Kurzweil, as we have already discussed (see page 84).
According to Google's top brass, artificial intelligence
will by that date be a billion times more powerful than
every human brain combined. The last phase, as has
now been revealed by certain members of Google's
management team, will be an interface between artificial
intelligence and our brains. Even Elon Musk has now
joined in this fascination for cerebral augmentation,
declaring at the website Recode's Code conference on

2 June 2016 that the survival of the human species in the face of AI will require us to interface our neurones with electronic components as soon as possible!

Jean-Michel Besnier: You keep coming back to Google and the Silicon Valley companies, but I think the problem is more complex. Any company that believes it will owe its prosperity to technological innovation will cheerfully dive into the slipstream of the astounding pronouncements that have become a speciality of the transhumanists. Innovation culture is first and foremost a decision to entrust to the market the burden of deciding if a product deserves to survive and be developed. Such a product will not have been dreamed up to meet a specific need identified by programmers but instead will have more of its impetus in an idea originating from an engineer, a designer, a financier, an industrialist or whatever – an impetus that will result in that product's being selected (or not) from the wide range available for consumption. It is this quasi neo-Darwinian genesis that serves as a guiding principle for such an economy, which no longer sees itself as a "moral and political science" (to borrow a phrase from the economist Albert Hirschman), nor the simple expression of an invisible hand, but as a hyper-liberal or libertarian religion, waiting for the Great Machine to grant innovations their chance to enrich their audacious creators. In this sense, the startups now trying hard to make more and more apps are characteristic of the enterprises serving the cause of transhumanism: launching onto the market innovative products that will multiply across the board, before creating conditions for change that will herald the rise, if not of the posthuman, then of a human race at one with its machines. That same information economy is the kingdom in which the prophecies of the Singularists inspired by Kurzweil will flourish. Creativity is measured in terms of inventing intelligent objects, unusual interfaces, digital platforms intended to make our lives easier – in other words, a more connected reality. The French website *Soon Soon Soon*, assisted by numerous "lifestyle" researchers scattered around the planet, has specialized in collating the widest range of gadgets that herald the next transformation of our daily lives through technical invention. Health has often been fertile ground for these new ideas. Essentially, people are ready to incorporate anything into their daily lives in order to live longer – from including fitness trackers and food scanners to remote diagnosis via the internet, cheap DNA sequencing, and nanorobots injected in capsule form.

Is it bad, Dr Google?

By 2030, there will be no medical diagnosis without recourse to an expert system; there will be a million times more data in an individual's medical records than is the case today. This revolution is the result of parallel developments in genomics, neurosciences and the internet of things. A complete biological analysis of a tumour, for example, amounts to 20 trillion bits of information. Numerous electronic sensors will soon be able to monitor our health; objects connected to the net, such as Google's smart contact lenses for diabetics, will therefore produce thousands and then billions of bits of data for every patient, every day. Google X, the company's semi-secret laboratory, is perfecting a system for the extremely early detection of illness using nanoparticles, which will also generate an astronomical amount of data. Doctors will be faced with a veritable "digital downpour" and will have to interpret millions and billions of bits of information, where now they deal with no more than a few handfuls of data; even Dr House, from the eponymous television series, would be unable to cope with such an avalanche of intelligence. Can the profession adapt to such an abrupt transformation? The reality is that Watson, IBM's expert system, is capable of analysing hundreds of thousands of scientific studies to gain an understanding of a cancerous mutation in a few seconds where an oncologist would need 38 years, working night and day, to deal with a single patient; this is more than the life expectancy of the patient, and even of the oncologist. As there is no chance that a doctor can verify the millions and billions of bits of data that medicine will produce, we are going to witness a radical and painful transformation in medical power; doctors will be signing prescriptions they have not personally specified. There is a considerable chance that physicians will become the nurses of 2030: subordinate to algorithms, just as a nurse is today to a doctor. Another side effect will be that medical ethics will no longer be the explicit product of the doctor's brain; it will be generated more or less implicitly by the expert system. Medical and ethical power will be in the hands of designers and software engineers, and such expert systems will have monstrous power and intelligence. The leading lights in the digital economy – Google, Apple, Facebook, Amazon (GAFA), not to mention IBM and Microsoft – will doubtless become the masters of this new kind of medicine. LA

Laurent: All that is a little bit anecdotal. Nonetheless, in a few decades Google will have transformed humanity – what was once a search engine will become a neuroprosthesis. "In about fifteen years, Google will answer your questions before you have even asked them. Google will know you better than your intimate partner does. Better, probably, than even yourself," as Ray Kurzweil has proudly stated. He, too, believes that we will be able to transfer our memories and our consciousness to microprocessors by 2045, allowing our minds to survive our biological deaths. IT and neurology will be as one!

Jean-Michel: We shall see. While we're waiting, it's obvious that any company that makes money with data has an interest in following suit with the transhumanists. And what company can resist total information? Joining up data, developing data networks, turning users into products and exploiting the data they have consented (or not) to give up…GAFA lead the field here, followed by all those calling themselves social networks, insurance companies, derivative funds, search engines or online shops. In his book *Who Owns the Future?* (2013), Jaron Lanier, one of the Californian pioneers of digital technology, describes "Siren Servers" that proliferate on the web, hoovering up "big data" – often without paying for it – and requiring more and more artificial intelligence to process it as they stray ever closer to the catastrophic scenarios pictured by the transhumanists. Managing and harnessing the immaterial resources that make up the information economy is, of course, fuel to stoke the fire of what has been called cognitive capitalism, explaining the rise to power of financial structures (think companies like Enron, Long-Term Capital Management...) that reinforce the impression that we have lost all initiative, on which the transhumanists base their argument to prepare us for the Singularity. It is astonishing that economists do not predict very much anymore, abandoning prospective reflection and planning in favour of modelling complex systems in which we are immersed in deregulated markets and globalization. Put quite simply, the world of algorithms reigns supreme, and decision-making (for example, whether to invest in the market or pull money out) is undertaken only on the statistical correlations supplied by databases that owe more to AI than to human grey matter. Here, too, the economic activity of technologized communities demonstrates its accordance with the transhumanist vision of a world in which humans are no longer the future. The "humanistic information economy" advocated by Jaron Lanier seems somewhat utopian.

Approaching the end of money?

In our meritocratic societies, it is principally differences in mental ability that (rightly or wrongly) determine differences in income or capital. This template will be shattered by AI, however. Human intelligence will eventually become risible in comparison with the abilities of machines; will we accept enormous disparities in income in this brave new world? If we take on board the intracerebral prostheses suggested by Google management, meaning that our performance is linked to the power of the implants in our brains and not to our intrinsic abilities, what will differences in income between people be based on? What's more, an AI society could become a society without work, which would deprive money of its function. If a billion cancer researchers can be emulated on a variety of hard disks in a matter of seconds, for example, what value will we attach to a human oncologist? It will be possible to design and produce all goods and services on a machine in an infinitely more effective way than when carried out by any human, even an augmented one. The meritocratic system will go up in smoke; how can the allocation of capital be organized if the concept of merit is impossible? The best solution will no doubt be the equal distribution of goods and services to each individual – Communism 2.0, where each person will receive according to his or her needs and not according to the amount of work done. It will be AI, and not the economist Thomas Piketty, the author of the global bestseller *Capital in the Twenty-First Century* (2013), that will put an end to inequality of income; capitalism will not survive intelligent machines. LA

Laurent: I agree entirely. Neurotechnology is literally revolutionary in that it overturns social order. Will we be able to escape this? Will a "counter-neurorevolution" be possible? Probably not; ultimately, a human being who refused to be hybridized with electronic circuits would hardly be competitive on the labour market. Can we imagine a two-speed society, with unaugmented humans who would inevitably become pariahs? And would it be ethical not to augment the cognitive abilities of less gifted individuals? Bill Gates himself is distraught at the lack of political discussion about the consequences of combining artificial intelligence and robotics. He estimates that machines will have replaced the majority of jobs, including in the healthcare professions, by 2035. The increasing power of neurotechnology is even ruffling a few feathers within Google, which has recently created an ethics committee dedicated to artificial intelligence. It is to deliberate on the questions that will affect humanity as a whole: should any limits be placed on artificial intelligence? How can it be controlled? Do we have to allow it to interface with our biological brains? The head of AI at Facebook has sought to offer some reassurance: in the 17 June 2016 edition of France's *Figaro* newspaper he stated that a *Terminator*-style scenario is out of the question for the next 20 years or so. But 20 years is nothing! In an age of brain prostheses, the risk of neuro-manipulation – *neuro-hacking* – and neuro-dictatorship is immense. We have to circumscribe the power of the neuro-revolutionaries; control of our own brains will become the primary entitlement among the rights of Man.

10
DO WE NEED TO
PASS SOME LAWS?

Is the upheaval augured by new technology inevitable? What can the government do? Is it possible to instigate – as a matter of urgency – a technological democracy that would be in a position collectively to face up to the immense challenges posed by transhumanism?

Laurent Alexandre: Rarely has humanity been confronted with such enormous challenges; controlling our fate over the long term is becoming our most pressing political task. The revolution that is set to radically change our civilization is being fomented along the shores of the Pacific at the behest of the digital giants and the leaders of China. Having colonized the cyber world, GAFA (Google, Apple, Facebook, Amazon) are moving into strong positions in robotics, artificial intelligence, genetics and nanotechnology. The Chinese leadership, trained engineers, have been actively promoting the China 2050 agenda. Once masters of their own fates, our governments are paralysed in the face of these new economic actors, who are inventing the future; technological illiteracy is the rule among state leaders. The futurist Joël de Rosnay rightly claims that to understand the future, you must love it. Meteoric NBIC technology would justify a reinvention of the regulatory role of the state. Technology and legislation will fuse, and "code is law" will become a political *fait accompli*. In relinquishing its supervisory role, the state is shooting itself in the foot and allowing technology to shape society at an increasing pace. Without our noticing, the balance of power is shifting, as technology becomes stronger than the law. It is interesting to hear Peter Thiel, the co-founder of PayPal and financier of Silicon Valley, point out that "a great company is a conspiracy to change the world". High-tech firms intend to play a political role!

Jean-Michel Besnier: For me, the principal problem is not the state, but democracy, so I would reformulate the question we are addressing as follows: how should we manage the control over humanity that we have acquired through advances in research? This refers to the objectives of bioethics but also *a fortiori* addresses politics (if we envisage it, ideally, as a system within which the decisions concerning the conditions for our mutual coexistence are taken and mediated). So let's recognize that technology is *a priori* incapable of imposing limits upon itself – we are essentially saying that it is where the immoderate behaviour (what the Ancient Greeks called hubris) that humans are capable of is expressed. A brake can only be applied from outside – it has to be mitigated by the product of thought and symbolic representation (the political communication permitted by language).

An ethical committee for NBIC?

In considering the notion of ethical regulation of technological innovations, we might take a look back at the precedent of biotechnology. Under President François Mitterrand, France in 1983 convened a

National Ethical Consultative Committee (the *Comité consultatif national d'éthique*, or CCNE), tasked with raising awareness in modern society of bioethical problems. (An annual public meeting combined with regular published updates has been a way of testing how the message is getting through.) The consultative opinions published by the CCNE are not destined to be enforced by law as such, but they feed into the work of legislative bodies; bioethical law (which has recently undergone great reform) is largely inspired by these. In that they reflect the concerns engendered by biomedical technology (for example, medical imaging, transgenesis and stem cells), these laws appear to derive from regulation imposed on ethical grounds. That is to say that the weight accorded to the so-called anthropotechnology seems to arise from the deliberations of wise counsel, in ways sanctioned by the political authorities, such as the Council of Ministers and the President of the Republic. JMB

Taking things a little further, and in order to avoid the undue professionalization of ethics, we should set our sights on a technological democracy, like the one advocated by the sociologists of science Michel Callon and Bruno Latour, for example. It would mediate the representative democracy that has been guilty of hardening the opposition between decision-makers and users or between experts and amateurs. It would allow itself to be constrained by the decisions of government alone but would extend the field of expertise in ethical matters to a range of people involved in biotechnological developments (patients, carers, researchers, industrialists, engineers). The methods applied to ensure that this was discussed and evaluated by society and that innovations were subject to regulatory control might consist of hybrid forums or town hall meetings with the public, for example. The challenge is to "politicize" the technology used to augment humans by framing it within a discussion and requiring that it be subject to democratic arbitration.

The game is far from won; think of the EU's decision, undiscussed by the population at large, to finance (to the tune of 1.019 billion Euros) the Geneva-based artificial intelligence programme known as the Human Brain Project. The odds are pretty good that this decision would not have gone through had the scheme undergone joint scrutiny by the scientific community and a cohort of "informed lay persons". Some research organizations such as INRA, Europe's top agricultural research institute, have instituted "pre-research programme consultations" that involve user associations and panels of non-experts. The idea is a good one, and expedient to avoid the log jams that governments run up against when claiming to be organizing public debate to gain acceptability for technology.

Laurent: Unfortunately, the political system is governed by emotion and pressure from the media, undermining the legitimacy of the state, which traditionally lies in an accommodation of the long-term interests of society. It is only logical that political impotence results in increased interest in authoritarianism. In 1995, according to the World Values Survey, less than 20 percent of the wealthiest Americans said they would approve of "having a strong leader who doesn't have to bother with Congress or elections", and 20 years later more than 40 percent said they would approve of it.

Jean-Michel: And what if that authoritarian power decided to improve humanity? To augment it? It's a valid question. Will it be done so that augmented humans are more effective in the rat race of our society – so that they can keep pace with the machines that we have allowed to multiply, as in the eugenics advocated by the British scientist Francis Galton at the turn of the 20th century? Will it be done because it is becoming more and more difficult to live as a vulnerable being doomed to suffer disease and die? In any case, such an "augmented human" (semantically conflated with an "improved human") is not an inevitability, and there is no lack of opposition to his advent. It will split humanity between those who will benefit from technological additions and those who will lack the means to access them. It will risk limiting the perfectibility and normativity inherent in the biological organisms that we are. It exposes us to the extinction of free will in favour of dehumanizing, technoscientific determinism; it clears the way for cloning programmes that will render it possible to transfer one human to the genome of another... The regulation of technology that improves humans will not obviate scrutiny of these socio-anthropological consequences. Nor can it shirk the philosophical questions, questions that the average citizen is more and more inclined to accept: should we really be wishing to eliminate chance from the human condition? Will controlling evolution not eventually be fatal, if we are thinking about putting an end to diversity and hybridization of life? Will we manufacture autonomous beings, subjecting them to external formats and norms?

Laurent: I'm with you on the questions that you raise, but you can see that public authorities are incapable of discussing them, even though they are the most pressing issues of the day. The state is dumbfounded in the face of the floods of development coming from Silicon Valley and yet dawdles along at the pace of the average senator. Recent political debate has been pathetic, given what is at stake. Democratic leadership must be reformed as a matter of urgency – it has been taken hostage by the tyranny of short-termism, which has proved incapable of understanding the NBIC revolution. Will it be possible to retransform politics through digitalization before our fate is sealed by technological cabals and the foundations of their ultra-rich owners, not to mention enlightened dictatorship, which invariably thinks in units of a thousand years? Or should we instead fear that e-politics will support the rule of immediacy and choke the life from any long-term vision?

Digital philanthropy

The fragmentation of political power has been thrown into even sharper relief with the emergence of a third agency professing to take an extremely long view – philanthrocapitalism. This combines the professionalism of the great captains of industry with a messianic vision espousing medicine and science. Bill Gates (the co-founder of Microsoft) and the businessman Warren Buffett have disinherited their children in favour of bringing about vaccination cover in Africa that had previously been thought impossible. Paul Allen,

Microsoft's co-founder, has industrialized brain genetics. In December 2015, Mark Zuckerberg, the founder of Facebook, announced that he would devote 99 percent of his fortune to promoting individualized education, medical innovation and social equality. Elon Musk, the founder of SpaceX, which has revolutionized access to space, has thrown open the doors of a foundation whose mission is to develop artificial intelligence. All these examples show that the big beasts of Silicon Valley are taking the long view and are ready to devote a good portion of their wealth to ensuring that the ideas dear to them – including transhumanism – see the light of day. LA

Jean-Michel: There is unfortunately a risk that your depiction of a destitute political world will attract sympathy. However, any such sympathy will soon give way to apprehension as the realization dawns concerning the totalitarian potential inherent in e-politics – a policy that champions technological methods and dispenses with those of the people is certainly undesirable. Worrying trends are already gaining in popularity among young people: we learn, for example, that 4.5 million young Americans take the stimulant Ritalin to boost the ability to concentrate needed for their school studies. The parents seem untroubled by the potentially harmful side effects, which reveals that they no longer expect education to deliver the power to accomplish what an amphetamine supposedly does mechanically. The interpretation of "improvement" here is in purely technological terms and not the result of an educational process. As long as there is an appetite to regulate the consumption of products for technological augmentation of human beings, we may be able to keep a lid on the deficit in the symbolic function borne by the intention to educate – on which the preservation of what is human depends.

Laurent: I agree, the Scholars Academy will play a key role in preparing for what is taking shape before our very eyes. In parallel with the search for a framework for AI, a deep rethinking of the role of the education system is essential. Having more or less failed to evolve in 250 years, education today is essentially as behind the times as medicine was in 1750. Its organization, its structures and its methodology are frozen in time, and – worse still – the university system is training people for outdated jobs. How are you supposed to teach children who are going to be growing up in a world where intelligence will know no limits? Hitherto, every technological revolution has involved a transfer of jobs from one sector to another – from agriculture to industry, for example. With AI, however, there is a considerable risk that a lot of jobs will simply be wiped out, rather than transferred. What should we pass on to children to ensure that they will have a well-rounded education in this brave new world? As far as transfer of skills and training for life is concerned, the education system in its current form is already outdated technology. Schools in 2050 will no longer deal in knowledge but in brains, thanks to the fields of NBIC. The three pillars of the Academy – content, methodology and staff – must therefore be rethought. For a start, the humanities and culture in general need to be rebooted, as wishing to compete with machines on technological matters will soon be a risible enterprise. We will then have to tailor teaching to the neurobiological and cognitive characteristics of each student. Last, we will have to bring specialist neuroscientists into the education system, as teaching will essentially be centred on neuroculture by 2050. The introduction of NBIC to improve education techniques will also require a parallel in-depth rethinking of neuroethics: nobody wants schools to become places of neuro-manipulation. Given the challenge presented by the ubiquity of AI, we must embark with all urgency on the modernizing of education. It is the only way we can give the lie to the prediction made by Bill Joy, the co-founder of Sun Microsystems, who announced in the year 2000 that "the future doesn't need us".

11
HAVE WE ANYTHING TO FEAR FROM A "BRAVE NEW WORLD"?

We all know Aldous Huxley's 1931 science fiction novel *Brave New World*, in which he describes a totalitarian world where the state has seized the right to select which babies are destined to live, consigning them to a caste according to their biological potential. Given the rapid evolution of technology, however, Huxley's dystopia could soon become reality. How should we deal with it?

Laurent Alexandre: How can we make predictions about the 21st century, which will witness greater change in humans than in the last few millennia? Our minds are not ready to think so far ahead, and so we either idealize the future or paint an excessively gloomy picture. Our initial reaction is to predict the demise of humanity – that we are going to enter a universe that is, say, ice-cold, hostile, dehumanized and run by mad scientists. According to bioconservatives, there will be nothing human left in a posthuman, stuffed full of electronic components. Bioconservatives feel instinctively that the future looming ahead appears to go against nature.

Jean-Michel Besnier: Any democracy is in danger of spinning out of control, into either anarchy or tyranny – a gloomy prophecy that dates back to Plato. The political theorist Alexis de Tocqueville was no more encouraging in the 19th century, venturing a prediction that our age seems to be fulfilling: that democracy will produce individuals who become increasingly fearful of their liberty and of the instability in the world that such freedom propagates; they will feel more fragile and insecure. The result will be that they demand ever greater intervention from the state (the welfare state and, unwittingly, the surveillance state) and will instigate a soft despotism that will relieve them of the worry of being free. The totalitarian regimes of the 20th century unfortunately exploited the slippery slope on which democracies find themselves when they neglect to take precautions against their members' failure to assume political responsibility. They went too far, however, in claiming to create from first principles a new man who would be divorced from history, but who would then believe he had attained perfection. We thought we had saved ourselves from these totalitarian ideologies and were able to accept the notion that democracy is "the historical regime *par excellence*", as the philosopher Claude Lefort put it, and this entails our taking risks and standing up for our values (an essential in itself). We imagined that Aldous Huxley's *Brave New World* and George Orwell's *1984* had cured us of the illusion of unbearable happiness. Oh no, we thought, never again would we allow a world to arise that would manipulate people and keep them docile via propaganda and controlled amnesia. But we had not anticipated that, while the prospect of political revolution fortunately no longer appealed to anyone (of whatever partisan stripe), science and technology might take up the baton and announce pure happiness in the nearest future, or indeed have it brought about by some posthuman or other.

Laurent: That's an entirely classic aspect of history – you never see technological revolutions coming.

The notions of transhumanism and posthumanism we are discussing might seem like science fiction straight out of Hollywood – especially in Europe, where there is still far less awareness of the progress to come from the confluence of the NBIC sciences. To the average citizen, the idea that we might be a few decades away from a posthuman world populated by hybrid individuals would seem like an outlandish millenarian theory, but we would do well to remember that "serious people" have mocked utopian thinking in every era.

An anthology of blindness to progress

The renowned University of Chicago astronomer Forest Ray Moulton said in 1932: "There is no hope of ever reaching the Moon one day; it is physically impossible. There are insurmountable barriers to escaping the Earth's gravity!" In 1956, the British astronomer Sir Richard Woolley, who became Astronomer Royal that year, declared: "All this talk about space travel is utter bilge!" The next year, the highly respected American engineer Lee De Forest put the last nail in the coffin when he announced: "To place a man in a multi-stage rocket and project him into the controlling gravitational field of the moon where the passengers can make scientific observations, perhaps land alive, and then return to earth

– all that constitutes a wild dream worthy of Jules Verne. I am bold enough to say that such a voyage will never occur regardless of all future advances!" Within four years, in 1961, Yuri Gagarin of the Soviet Union was floating in space, and eight years after that, the American astronaut Neil Armstrong had set foot on the moon. Similarly, even the most brilliant geneticists underestimated the genetic revolution. In 1970, Jacques Monod, who won a Nobel Prize in Medicine for discovering messenger RNA, wrote in his *Chance and Necessity* (1970): "The microscopic scale of the genome will no doubt always prevent it from being manipulated." Scarcely five years later, the first attempts at genetic editing had been successfully completed! And as for DNA sequencing, just 30 years ago biologists at the top of their field were claiming that we would never be able to sequence our chromosomes in their entirety or would have to wait at least until the years 2300 or 2500 to do so! It had been achieved by 2003 and we can all expect to be sequenced by 2025. This underestimation of technological progress prompted Wernher von Braun, the father of the US space programme, to state: "I have learned to use the word 'impossible' with the greatest caution." LA

Jean-Michel: This argument about our blindness to the future is not without merit as far as technology is concerned, but all it does is prevent dystopias from dying; they have merely changed tack and now claim to be based on facts established by scientists (the findings offered by neuroscientists, for example) or an extension of the realities promised by technologists (the circumstances imposed by digital globalization, for instance). Far from having learned our lesson from totalitarian regimes, we are witnessing a growing enthusiasm for declarations in favour of manufacturing a human being, of manipulating his genome and his moods, of suppressing his existential fears and the chance that sharpens them, of the reasons to believe that his immortality might be conceivable. Decoded in the light of the criticism of totalitarianism from Raymond Aron, Claude Lefort, Cornelius Castoriadis, Marcel Gauchet and even Hannah Arendt, the promises of renewal professed by the transhumanists contain elements that would dismay even the most credulous. The deaf ear turned to philosophical analysis (such as that proffered by Michel Foucault on biopower or Gilles Deleuze on the "society of control") is truly heartbreaking and would be reason enough to opt from now on for consumption of NBIC without compunction. The totalitarianism that threatens us will be of the order of the fascism described by the semiologist Roland Barthes: it will forbid us less than it compels us to act, pleading the guarantee of biotechnology, and will function on the basis of discrediting politics and the certainties imposed upon all as evidence that science and technology can offer conditions in which people will thrive individually and collectively – in other words, the satisfaction (or the opposite) of the desire that is fuelled by absence. The promised happiness might therefore be thought of in terms of social insects, entirely interlinked and producing a painless homeostasis. (I am thinking of the collective brain brought about through the interconnection of the neurones that make up us all, in the countless networks that we will create.)

Laurent: You mention the threat of bio-totalitarianism. I think that people will offer less resistance to the biotechnological revolution where it promises an advancement of their own power and victory over death. Given that society is developing at breakneck speed, who in 2080 will want to remain an obsolete human, fragile and sick, when his neighbours are "geniuses" and more or less immortal? Who will want to settle for a run-of-the-mill IQ and simple human memory when biochips can provide artificial intelligence superior to that of millions of aggregated human brains, with immediate access to every database? The crowd instinct, not to mention peer pressure and the need to stay within the norm, will guarantee compliance by the vast majority.

Jean-Michel: This question of "staying within the norm" seems a deciding factor to me. How can we not believe that transhumanism in the radical form you outline will not betray the "weariness of the self" (to borrow the title of the book by sociologist Alain Ehrenberg) and the depression of *Homo democraticus?* Of course, you might be tempted to object that modern citizens are not cut off from their peers, as individuals in totalitarian states were. You will say that they belong to as many social networks as they wish to and are therefore part of experiences that are propitious for the creation of a collective intelligence. The danger of transhumanism lies in the fact that it calls upon biotechnology to be the source of aspirations to create a posthuman being, without articulating any laws of history that are reliably falsifiable — just as Stalinism did. In this respect, its seductive power is all the more pernicious. With its ability to normalize at great speed — bringing humanity back to its only biological infrastructure — it makes total domination by technology inevitable, and therefore more surely lends credence to an idea that had sadly been substantiated by the totalitarian regimes of the past: humanity is a failed species that it is time to replace.

Democracy 2.0?

The renewal of democracy through digitalization is a refrain that is struck up by people from time to time – but who will truly join in on the chorus? Day in, day out, reality presents us with evidence that our identities are becoming nothing more than our digital trails, deploying no other content than the results of our wanderings on the web; it shows that the sense of being irreplaceable, the prerequisite of all morality, is no longer an issue for our time, no more than that of an assumed subjective responsibility; it means that any interior life will henceforth be an "anti-value" to be expiated or defeated. The world will be bio-totalitarian to the extent that it will permit the victory of the technoprogessive obsession with one's own survival, for longevity with no end(ing), for biological individuation shorn of the symbolic dimension that makes up human existence. You would have to be a pretty lightweight thinker (or cynical) to believe that democracy would succeed in drawing only on the arsenal of options presented by the web, such as hybrid forums, petition sites, blogs, tweets. However you slice it, *Homo communicans* will consent to existing only in transit, in a headlong rush forward, and will blithely conflate demagogic acclamation and deliberation, exhibitionism and confidence, transparency and authenticity. JMB

Laurent: You are arguing as if this replacement of humanity has to take place right across the planet, and I would like to reinforce one point here. A great number of intellectuals have called for the installation of a world government; I think it's claptrap. While the decision to create on earth some form of intelligence other than our own can be taken only at a global level, some areas should not be regulated in a centralized way; having a single decision–making centre could lead to a totalitarian system as no one would be able to escape it. It is essential to maintain a multiplicity of geopolitical power bases in order to guarantee ideological competition. The need for opposing powers and pluralism will be necessary for biopolitics just as it is for traditional politics today. You must have somewhere to run away to! Neurosecurity, and thus the protection of our brains, is a case in point; a world where regulation of brain science was decided at a global level would leave no escape routes. If neuroscience descended into totalitarianism, where could we seek exile? There would be nowhere left that was not subject to central neurobiotechnological power. This would indeed be a nightmare for our freedoms. On another note, we will clearly have to extend the Hippocratic oath to NBIC science experts, in particular to those who work in neurological science!

12
HOW FAR SHOULD
OUR RESEARCH GO?

*Thanks to the use of nanobots [microscopic robots] in our
brains, we will have almost God-like powers by the 2030s.*

Raymond Kurzweil, 2016

Man – obviously – lives in a universe. But
couldn't that universe itself be improved? The
prospect enthuses advocates of transhumanism
and horrifies its critics. Just how far should Man
push his ability to modify not only himself but the
world around him?

Laurent Alexandre: I am going to talk to you about the very long term, the distant future. What is going to become of humanity many years from now? Philosophers have always had a passionate interest in the origins of the universe; Leibniz's famous question ("why is there something rather than nothing?") dates back to 1697 . By contrast, very few thinkers have shown any interest in the future of our universe, yet the fate of our cosmos will be apocalyptic! The six scenarios modelled by astrophysicists, from the "Big Crunch" (a reverse Big Bang) to the "Big Chill" (the dissipation of all energy), all result in the death of the universe and the disappearance of any sign of our existence. Having only recently become aware of the need for sustainable development of our earth, we now learn that the universe itself is mortal.

Jean-Michel Besnier: Thinking about the distant future seems a pointless task to me, particularly as there is a great deal of uncertainty and our current problems are so pressing. Is it worthwhile to speculate on such distant time frames, billions of years in the future?

Laurent: On the other hand, I think that this leap into the future is of use, as it questions our values. Are good and evil relevant ideas on a cosmic scale? What meaning do our lives have if all trace of our civilization will disappear with the death of the universe? What is the ultimate goal of humanity, of science? A young French philosopher named Clément Vidal managed to sum up what is at stake in this programmed demise in his book *The Beginning and the End* (2014). For Vidal, the answer to the last question is clear: the ultimate aim of science is to combat the death of the cosmos by artificially manufacturing new universes. After the defeat of death, science will devote itself to fighting the death of the universe. Artificial cosmogenesis would mobilize every last scrap of humanity's energy over the next billion years. After regenerating our aging bodies with stem cells, the aim of our regeneration of the cosmos would be to make the universe immortal or replaceable.

Jean-Michel: Once again, I find such speculation over the very long term pointless. I am more interested in asking if there should ever be opposition to a limit to human perfectibility. Or to put it another way, should we ever refuse, at any particular moment, to pursue our quest for progress, at the risk of putting an end to human history? I myself have absolutely no sympathy for those archaic societies that demand we imitate old practices and observe traditions as an absolute norm intended to stave off the necessarily harmful effects of time. However, I don't think it is possible to debate modernist ideas, because if progress cannot be discussed, it would impossible to formulate values and guidelines to orientate our actions. We would have no choice other than to opt for facts, like the positivists. Because I am neither an animal nor a machine, I refuse to allow facts to be the law and science to dictate its formulation, while the moral perspective is left in limbo and ignored. That said, everything is yet to be explored. How can something that purports to be an improvement be properly scrutinized? By showing that in reality, it is really no such thing: because it entails collateral or perverse effects, because it doesn't serve the ends of all or because it is the flip side of manipulation or instrumentalization? It is not being reactionary to regard the improvement that is claimed to come about through NBIC as the implementation of a bias (whether ideological, political, industrial or social) in respect of the format that one would wish to impose on humanity – a bias that is per se up for discussion. I have already sufficiently laboured the point that I find the "integrated thinking" that would be the upshot of interlinking my brain with the web *a priori* nauseating; that smart medicine would seem to be a waiting room for mass hypochondria; and that I find the ultimate goal of achieving long life tempting, but on the condition that the ethical/political question of leading a good life is not brushed aside as insoluble. After all, when the fact that we now sleep for a much shorter time than in the 19th century is presented to me as progress, I may certainly tally up the extra time this has added to my existence, but will also add up the amount of stress that insomniacs like me have to suffer! You still meet people who don't want their living conditions to be improved through various devices, who want to retain the burden of deciding what is good for them. From their point of view, technology is even the negation of all wisdom, as it *a priori* excludes us from aspiring to discover our place in the world, by acting as if the cosmos had assigned it to us for all eternity.

Laurent: You mention wisdom. I myself am pondering religion, as the Singularity proclaimed by Kurzweil amounts to a new religion. This vision of the human of the future, omnipotent and immortal, is reminiscent of Hollywood scenarios like in the film *Transcendence* (2014) and is risible. It invariably expresses a basic shift: for the first time, a philosophical movement is claiming to rescue Man from his status as a creature at the mercy of nature and transcendence in order to give him an active role in evolution.

A new religion

Transhumanism is the final phase in the evolution of religious thought, which has gone through three stages. First, there was polytheism (the logical successor to shamanism), which reached its acme under the Romans and Greeks. This was followed by monotheism and the three Abrahamic religions. A third age is now emerging: that of the man–god. For transhumanists, Serge Gainsbourg's quip "Man created God. The inverse remains to be proven" goes without saying. God is yet to exist: He will be the man of tomorrow, endowed with almost infinite powers, thanks to NBIC. Man is going to bring about what only the gods were supposed to be able to achieve: creating life, modifying our genome, reprogramming our brains and putting an end to death. LA

Jean-Michel: As I have already said, as far as I'm concerned, boosting life and even killing death are absurd undertakings if they can't be "framed within culture" – in other words, placed within a symbolic dimension likely to make the human condition more desirable. A "transhumanism at the service of social progress", as advocated by Didier Coeurnelle and Marc Roux, the moving forces behind the French transhumanist association AFT Technoprog, might well wish to "improve our human understanding of communal life", as they say in their book *Technoprog* (2016). But how will it do this without negating the disproportionate role given to signifiers in our daily lives – without resisting the automatism that renders our existence subservient to technology? The question it raises would be all well and good: to what extent should we allow technological innovations to eliminate the risks and hazards of communal life in favour of supposed social harmony intended to protect us from self-destruction?

Laurent: Some philosophers, including Clément Vidal, have answered this question by asserting that there is no limit. For the transhumanists, it would be logical, and not the ultimate in vanity, to make the universe immortal in order to guarantee our own immortality. In reality, transhumanism mediates the interrelationships between our abilities and our beliefs – just as with polytheistic and monotheistic religions. A Promethean religion exalting the omnipotence of Man over the elements would be inconceivable before the triumph of NBIC. Current religions have every intention of helping us deal with our death – through faith – but none can help us eliminate it! For most transhumanists, NBIC will rob God of all credibility and replace Him with a human cyborg. Is the religion of technology in the process of replacing traditional religion? Will there be violent struggles, indeed religious strife, between transhumanists and bioconservatives, or a gentle transition? In fact, the first bridges seem to have been built between transhumanism and religion: the Dalai Lama is passionately interested in neurotheology and cerebral control of religious feeling. Might Buddhism be the intermediary religion to usher in the transhumanist era? This third religious age is freighted with psychoanalytical hazards. In an exciting conference paper at the Catholic University of Louvain, Belgium, in 1972, the psychoanalyst Jacques Lacan explained how death helps us to live and why life would be terrifying if it were infinite; when everything is possible, human beings are sent mad. Psychoanalysis has shown us the degree to which the absence of constraint is a source of distress and turmoil, and transhumanist ideology, which inflates our fantasies of omnipotence, is certainly a vector of psychiatric pathologies. The transhuman will live in the illusion of his untrammelled power, which is fatal for our psychic wellbeing. One thing is sure, psychiatry is a profession with a future!

Jean-Michel: You have just quoted Jacques Lacan, a great interpreter of meaning and of the signifiers that endow us with humanity. I will finish with an anecdote that highlights the neglect of the opposition of signs and signals in which I see the main danger threatening our era: at the Salon du Livre book fair in Paris in March 2016, the philosopher Michel Serres and I were invited to debate in public the graphic novel *Adulteland* (2014), created by a young Korean artist called Oh Yeong Jin. We were talking about the reality and the imagined existence of augmented human beings and what made them appealing and repellent in equal measure. Hoping to put the last nail in the coffin of technophobes of every stripe (among whom he no doubt included me), Michel Serres appealed to the audience, addressing them more or less as follows: "We are all augmented beings, particularly in this room, where the books that surround us act as augmentations for our minds, which would be as nothing without the writing that ensures they will go on developing for ever!" The argument invariably comes back to why the undertakings of the transhumanists wouldn't be of the same nature as those that allowed the revolution of writing to take hold and flourish? I ventured to suggest to Michel Serres that, to my mind, the digital revolution, rather than continuing the writing revolution, seemed to threaten it. Contemporary technology, I continued, subjects us to signs and signifiers that are ever more numerous and insistent, demanding faster and faster behavioural reactions. Books, on the other hand, involve us in long-term relationships in which the signs and signifiers invoke a dialogue with ourselves, with the author, with other readers – with the humanity within ourselves and others, in any case. And let us hope that this book will be a further contribution to this dialogue!

FURTHER READING

Books:

Alexandre, Laurent, *La mort de la mort*, J C Lattès, 2011.

Alexandre, Laurent, *La défaite du cancer*, J C Lattès, 2014.

Atlan, Henri, *L'Utérus artificiel*, Le Seuil, 2005.

Atlan, Monique and Roger-Pol Droit, *Humain*, Flammarion, 2012.

Besnier, Jean-Michel, *Demain les posthumains*: *Le futur a-t-il encore besoin de nous?* Fayard, 2010 and Pluriel, 2012.

Besnier, Jean-Michel, *L'Homme simplifié: Le syndrome de la touche étoile*, Fayard, 2012.

Besnier, Jean-Michel, *La sagesse ordinaire*, Le Pommier, 2016.

Dyens, Ollivier, *La condition inhumain*, Flammarion, 2008.

Férone, Geneviève and Jean-Didier Vincent, *Bienvenue en Transhumanie*, Grasset, 2011.

Ferry, Luc, *La révolution transhumaniste*, Plon, 2016.

Haraway, Donna, "A *Cyborg Manifesto: Science, Technology and Socialist—Feminism in the Late Twentieth Century*" (1984), in Haraway, Donna, *Simians, Cyborgs and Women: The Reinvention of Nature*, Routledge, 1991.

Kleinpeter, Édouard (dir.), *L'Humain augmenté,* CNRS, 2013.

Kurzweil, Ray and Terry Grossman, *Fantastic Voyage: Live Long Enough to Live Forever*, Rodale Books, 2004.

Kurzweil, Ray, *The Singularity is Near: When Humans Transcend Biology*, Viking, 2005.

Lecourt, Dominique, *Humain, posthumain*, PUF, 2003.

Thiel, Marie-Jo, *La santé augmentée: Réaliste ou totalitaire?*, Bayard, 2014.

Websites:

European Transhumanist Party:
www.transhumanistparty.eu

French national ethical consultative committee:
www.ccne-ethique.fr

French transhumanist association AFT Technoprog:
www.transhumanistes.com

Humanity+, a transhumanist association:
www.humanityplus.org

NeoHumanitas think tank:
www.neohumanitas.org

World Transhumanist Association:
www.transhumanism.org

Three informative sites dealing with scientific topics:

www.piecesetmaindoeuvre.com

www.up-magazine.info

sciences-critiques.fr

INDEX